RECENT ADVANCES IN INFORMATION TECHNOLOGY

Recent Advances in Information Technology

Editors

Waldemar Wójcik & Jan Sikora

Institute of Electronics and Information Technology, Lublin University of Technology, Lublin, Poland

CRC Press
Taylor & Francis Group
Boca Raton London New York Leiden

CRC Press is an imprint of the
Taylor & Francis Group, an **informa** business
A BALKEMA BOOK

CRC Press/Balkema is an imprint of the Taylor & Francis Group, an informa business

Typeset by V Publishing Solutions Pvt Ltd., Chennai, India
Printed and bound in Great Britain by CPI Group (UK) Ltd, Croydon, CR0 4YY

Published by: CRC Press/Balkema
Schipholweg 107C, 2316 XC Leiden, The Netherlands
e-mail: Pub.NL@taylorandfrancis.com
www.crcpress.com – www.taylorandfrancis.com

ISBN: 978-0-8153-7387-2 (Hbk)
ISBN: 978-1-351-24317-9 (eBook)

Table of contents

Preface

The main them of **Recent advances in Information Technology** is modern information technology, without which it is difficult to imagine progress in modern technologies and manufacturing processes. A wide range of topics is addressed, from game theory to advanced control problems. The individual chapters are separate, closed issues, which are the fruit of the cooperation of Ukrainian, Kazakh and Polish scientists. The editors and the authors hope that the book will strengthen friendship between our communities and nations.

Waldemar Wójcik
Jan Sikora
Lublin, March 2017

CHAPTER 1

Security estimates updating of asymmetric cryptosystems for modern computing

V.K. Zadiraka, A.M. Kudin, I.V. Shvidchenko & P.V. Seliukh
V.M. Glushkov Institute of Cybernetic of NAS of Ukraine, Kyiv, Ukraine

P. Komada
Lublin University of Technology, Lublin, Poland

A. Kalizhanova
al-Farabi Kazakh National University, Almaty, Kazakhstan

INTRODUCTION

Today RSA is still one of the most common and widespread asymmetric cryptosystems. A lot of cryptographic protocols that deal with encryption, digital signature, and distribution of key information are based on RSA scheme. The basis of scheme's security is a complex theoretical and numerical factorization problem: to find the prime divisors of large n.

The issue of cryptosystem's security in practice is associated with the task of correct choice of cryptosystem parameters. A recent research by a group of scientists led by Arjen Lenstra (2012) confirms the problem of RSA implementation, namely generation of parameters that will ensure the system's security in practice. From among 11.5 million of RSA key certificates under research there has been found more than 26000 vulnerable keys with size of 1024 bits and 10 keys with size of 2048 bits. In this case the vulnerability was to factorize the cryptosystem's module.

This vulnerability was implemented through the use of common divisors for modules, what is more, only a small amount of such common divisors was found as a result of key reissuing for the same owner. At most this situation is caused by the presence of the global problem of generating "qualitative" modules that are built with large prime numbers (Bernstein et al. 2013, Heninger et al. 2012), besides, the authors conclude that crossing to larger modules does not lead to the expected decrease of vulnerable keys number. The mechanisms for generating prime numbers that are used for constructing modules should be additionally studied.

Given the above we formulate the problems of effectiveness analysis of applying different computational models (including probabilistic, parallel and hybrid) for factorization and discover the existence of polynomial complexity algorithm of numbers factorization of special form; and as a result to clarify practical security of RSA cryptosystem.

We will discover the amount of "secure" RSA modules. If we find out the limited amount of such modules of defined size and this limitation turns out to be the polynomial of the module's size then there is a critical issue of RSA security in practice, namely if the cryptanalysis or RSA is P (by factoring the module).

Adi Shamir offered the research of RSA module break "cost" of certain length by constructing a specific factorization machine (Shamir 1999). This research was continued in Shamir, Tromer and Bernstein papers (Shamir et al. 2003, Bernstein 2001). This task is urgent now due to cloud computing and flexible computing architecture models (Programmable Logic Device (PLD)) rise. To research the impact of new computer (cloud computing, etc.) technology on complexity and RSA module break cost is the main idea of the article.

1.1 THE EFFECTIVENESS OF DIFFERENT COMPUTATIONAL MODELS FOR NUMBERS FACTORIZATION

Let us briefly look through the characteristics and applicability of sequential, parallel and probabilistic computational models. It should be noted that quantum calculations and their applicability to the cryptography problems, in particular factoring problem have gained a wide interest in recent years (Zadiraka et al. 2013). P.W. Shor (1999) has proved the polynomial complexity of factorization and discrete logarithm for this computational model. This gives an additional possibility to construct effective factor algorithm and to discover more precisely the connection between computational models and their practical implementation. This topic requires a separate discussion, so it is beyond the scope of this chapter.

The most studied sequential computational model uses determined operations performed one after another and the results of the previous calculations are used in the next step. Deterministic factoring algorithms are fully expressed through this model of computations. Today there are no practical ways to significantly improve the performance of such factoring algorithms for arbitrary input.

The extension for sequential model is probabilistic. It is a good framework for inherently probabilistic algorithms. There are two types of algorithms: Monte Carlo and Las Vegas. The algorithms of the first type give the answer that may be correct for some random sequences that are generated during the algorithm, but may be incorrect. The probability of false answers is reduced by repeated execution of the algorithm.

The Las Vegas algorithms respond correctly or finish with answer "don't know". The probability of getting an incorrect result is zero. Exactly the study of probabilistic factoring algorithms can lead to getting an effective performance of algorithms for numbers factoring of special form. This can happen due to the "nature" of Las Vegas algorithm to "split" input data by subsets with different estimates of algorithm's complexity.

Consider the algorithm which can be divided into blocks of operations, the result of which indirectly affects the further work, so the branches for independent computing that can be executed simultaneously are allowed. This arrangement of computing defines the parallel computational model. The application of parallel model for k separated branches is expected to give at best the k time's speedup of algorithm runtime (linear decrease computational complexity).

Unfortunately, in practice, to achieve such speedup for factoring algorithm is impossible because of the inability to efficiently use CPU time of each of k branches through the existing need for communication between processors.

The well-known Brent's review (Brent 1990) states that certain factor algorithms ECM (Elliptic Curve Method (Lenstra 1987) for example) are good-suited for parallelization that yields almost linear effect of decreasing computational complexity, but some algorithms (including rho-Pollard method (Pollard 1975)) can achieve this effect only in theory (Crandall 1999).

Recent research of Pollard's method (Crandall 1999, Koundinya 2013) shows that the linear effect can be achieved only for the small number of parallel branches. Note that there are two possible ways to parallelize the algorithm—to build various elements of one random sequence, or to look up different sequences. Namely, the application of the second approach and possibility of its effective improvement are discussed below.

Note that the mentioned above can be applied only for general factoring algorithms. It is clear that these estimates are the upper bounds for factorization the numbers of special form. Moreover, the study the effectiveness of factorization algorithms determines the so-called problem of "impact of algorithm optimization on the PC's structure". Or rather it can be defined as the problem of constructing an effective computational model (possibly hybrid) for number factoring and discovering its implementation on the base of programmable logic devices PLD, GPUs. This confirms the relevance of discovering the existence of time-efficient factoring algorithms of special numbers, some results are discussed below.

1.2 RHO-POLLARD ALGORITHM

Pollard's rho-method is a probabilistic algorithm that once made a significant breakthrough and still is not completely researched. The complexity of the algorithm does not depend on the factor but on its divisors. Thus, like the ECM algorithm (Lenstra 1987), Pollard's method is effective for factoring multi-prime RSA modules, such numbers of fixed length that are the product of more than two primes.

Pollard used two simple ideas: the birthday paradox (Sekey 1993) and the ability to check divisibility of number n by p computing $GDC(x-y,n)$, if $x \equiv y \bmod p$. We can choose the starting point x_0, function f and construct recurrent sequence $x_i = f(x_{i-1})$ which plays the role of a "random" sequence. Considering it by the module p, where p is a divisor of n there must exist two comparable elements.

Pollard chose $f(x) = x^2 + 1$ as a recurrence. The subtraction of two elements is always split into a product $x_k - x_j = x_{k-1}{}^2 - x_{j-1}{}^2 = (x_{k-1} - x_{j-1})(x_{k-1} + x_{j-1})$. These divisors are tested as possible for the number n.

This brings up the question—is it possible to obtain a significant improvement in the performance of the algorithm by choosing another recurrent function. The study of this issue (Hartman 1982) was carried out only with the use of numerous assumptions that narrow the set of numbers to factor. Despite the application of different functions the effectiveness turned out to be negative. So we considered the sets of numbers reduced by the module p and the capacity of such sets for the following cases: $p = 2p' + 1$ where p' is prime; $p = 3(\bmod 4)$ or $p = 2(\bmod 3)$.

Hartman obtained the following results about the capacity of the sets that are generated by different functions.

1. $\{(ax+b) \bmod p, x = 0,1,\ldots,p-1\}$ is complete set of residues. It is determined as a permutation.
2. If k is odd and $(k, p-1) = 1$, then $\{x^k (\bmod p), x = 0,1,\ldots,p-1\}$ is a permutation.
3. If $P(x)(\bmod p)$ is a permutation, then $(aP(x)+b)(\bmod p)$ is a permutation for $a \neq 0 \bmod p$.
4. If $P(x)(\bmod p)$ generates the set of size n, then $(aP(x)+b)(\bmod p)$ and $P(ax+b)(\bmod p)$ are the different sets, but also of the size n.
5. $P(x) = x^2 (\bmod p)$ generates the set of size $\frac{p+1}{2}$. This statement is correct for all polynomials of degree 2.
6. If $P_1(x), P_2(x)$ are permutations, then $P_1(x), P_2(x)$ does not generate the whole set of residue. Little is known about the size of the set.
7. Cubic polynomial $P(x) = ax^3 + 3bx^2 + 3cx + d$ separates the complete set of residues into two size-equivalent subsets corresponding to the following:
   a. x^3 of size $p(b^2 = ac)$;
   b. $x^3 + x, x^3 - x$ of size $2\frac{p+1}{3} - 1(b^2 \neq ac)$.
8. $P_1(P_2(x))$ generates the set of size less than or equal $\min(P_1(x), P_2(x))$.

But to discuss only the size of set does not make sense. It is important to consider how much calculations should be done to find an item that is congruent to module p with the previously calculated one. If the selected function, and therefore, the generated sequence has many "small" loops and does not have "long" loops and "tails" then the probability to find quickly the decomposition of number n is high (Fig. 1.1). But hypothetically this is possible only under the condition that n has only small prime divisors. If sequence has "long" loops then the probability that the selected starting element will lead us to the desired collision is high and the probability of choosing a starting element, which quickly leads to collision, is low.

To use high-degree functions is impractical because of the large time-costs of sequence calculations. Using the linear function $f(x) = ax + b \bmod N$ the subtraction

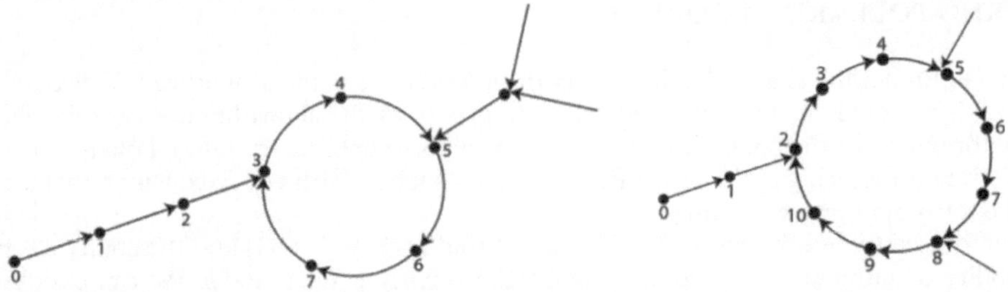

Figure 1.1. Loops and "tails" in rho-Pollard method.

$x_k - x_j = (x_0 + kc) - (x_0 + jc)$ does not seem to be random comparing with $f(x) = x^2 + 1$ proposed by Pollard (1975).

Applying two different linear recurring sequences also does not give any improvements in performance; we can conclude nothing about the residue of elements of two such sequences $x_k - x_j = kc_1 - jc_2$, where c_1, c_2 are the constants. As we can see, the matter of selecting the function that generates random sequence for Pollard's rho-method is still open.

1.3 SELECTING CRYPTO PARAMETERS

Let us define n as a module and e as an exponent for RSA cryptosystem; these parameters are the public key. Cryptosystem's module is a product of odd primes p_i, $i = 1,2,..,u$ where $u \geq 2$, and public exponent e is an integer that takes the value between 3 and n – 1, $GCD(e, \lambda(n)) = 1$, where $\lambda(n) = GCD(p_1 - 1, ..., p_u - 1)$ – generalized Euler function and some additional conditions (Mukhachev et al. 2005). Private key d is a positive integer that satisfies $e \cdot d \equiv 1(\mathrm{mod}\ \lambda(n))$.

Let consider RSA key size as a length of module n in bits. NIST limits the set of key size values as 1024, 2048 and 3072 bits. In addition, restrictions are imposed on the generation of module n only as a product of two primes $n = p \cdot q$ (FIPS PUB 186-4-1).

The Public-Key Cryptography Standards (PKCS) #1 of version 2.1 and more it is allowed to use modules that are the product of more than two numbers (multi-prime modules), but it does not determine any restrictions on the choice of divisors. For a fixed size of the module the usage of more than two divisors reduces the size of such divisors and thus increases the probability of these modules to be factored by the algorithm, the complexity of which depends not on the size of factored number but on the size of number's divisor.

To deploy RSA cryptosystem first we need to select exponent e, fixed or random integer $2^{16} < e < 2^{256}$. It is not permitted to choose random primes p and q to generate module with size of 1024 bits, but only those that satisfy the following (FIPS PUB 186-4-1):

- $p - 1$ has prime divisor p_1,
- $p + 1$ has prime divisor p_2,
- $q - 1$ has prime divisor q_1,
- $q + 1$ has prime divisor q_2.

Here p_1, p_2, q_1, q_2 are so-called auxiliary primes, p i q are primes with conditions. For modules of size 2048 and 3072 bits we may use random primes p, q, that are built using probabilistic or constructive algorithms of generating prime numbers (Maurer 1990, Gordon 1984, Zadiraka et al. 2007, Shawe-Taylor 1986). Table 1.1 shows the limitation on the length of numbers when using different algorithms of generating primes. The function $len(p)$ is the bit length of p.

Table 1.1. Valid length values for auxiliary primes p_1, p_2, q_1, q_2.

		Max. value of $len(p_1)+len(p_2)$	
len(*p*)	Min. length of auxiliary primes	*p*, *q* probable primes	*p*, *q* provable primes
1024	>100 bits	<496 bits	<239 bits
2048	>140 bits	<1007 bits	<494 bits
3072	>170 bits	<1518 bits	<750 bits

Table 1.2. Minimum number of rounds of Miller-Rabin testing when generating primes for RSA scheme, the error level is 2^{-100}.

Parameters	Number of rounds
$p_1,p_2,q_1,q_2 > 100$ bits p and q : 512 bits	For p_1,p_2,q_1,q_2 : 38 For p i q : 7
$p_1,p_2,q_1,q_2 > 140$ bits p and q : 1024 bits	For p_1,p_2,q_1,q_2 : 32 For p i q : 4
$p_1,p_2,q_1,q_2 > 170$ bits p and q : 1536 bits	For p_1,p_2,q_1,q_2 : 27 For p i q : 3

Probably the prime number p is a number generated by the following algorithm: to determine the length of the desired number p; to generate odd random number; to check if the number is prime using a probabilistic test. The most widespread is Miller-Rabin test. The algorithm tests if given N is prime following the next steps:

1. To calculate t and s so that $N-1=2^t s$, s has to be odd.
2. To randomly choose b, $2 \le b \le N-2$.
3. To calculate $y = b^s \bmod N$.
4. If $y \ne 1$ and $y \ne N-1$, then while $i<t$ and $y \ne N-1$ starting from $i=1$ do: $y = y^2 \bmod N$. If $y=1$, then Exit with "N is composite", otherwise $i=i+1$.
5. If $y \ne N-1$, then Exit with "N is composite".
6. Exit with "N is composite".

While constructing the module the following requirements on the amount of rounds of Miller-Rabin testing of the generated number should be met. This is caused by the decreasing probability to choose composite number as a prime with each additional round. Below there is the table with the amount of required rounds of testing for error level less than 2^{-100} is given.

Note that the use of probably prime numbers while implementing the cryptosystem not only significantly increases the time of module constructing but in case of mistaken use of composite number can cause malfunctions, which reduces the resistance to cryptanalysis. So we refer to deterministic algorithms that construct prime numbers in polynomial time. With the help of these algorithms we can reduce the problem of selecting the module by testing number's primality to the problem of generating provably prime number. Let's take a look at Maurer's (1990) and Shawe-Taylor's algorithms (1986). They are the so-called "tornado" algorithms: small prime number is used at the first iteration. Each following iteration determines provably prime number in a certain range that guarantees an increase of the resulting number.

Maurer's algorithm of generation of provably prime random number

The input of the algorithm is integer k – the number of bits of the desired prime. The algorithm uses the following parameters: L is the border for trial divisions; M is a parameter that ensures the existence of the desired prime. $M = 20$ is recommended.

1. If $2^k < L$, then to generate random odd k-bit number N and to test its primality with trial division. Repeat until a prime number N is generated. Exit.
2. If $k \le 2M$, then let $r = \frac{1}{2}$, else repeat the following until $k - rk > M$.
   2.1.1. Choose random integer s, $0 \le s \le 1$, $r = 2^{s-1}$.
3. $k_1 = [rk] + 1$. Repeat steps 1–2 to construct k_1-bit random prime q.
4. Let $t = 2^{k-1}/(2q)$.
5. Let R is random integer, $t < R \le 2t$. $N = 2Rq + 1$.
6. Let a is integer and $1 < a < N - 1$. If $a^{N-1}(\bmod N) = 1$ and $GCD(a^{2R} - 1, N) = 1$, then Exit with "N is prime".
7. Else repeat steps 5–6.

The second algorithm of constructing provably prime random number is Shawe-Taylor's algorithm. The input of the algorithm is an integer k – the number of bits of the desired prime. The algorithm uses parameter L as a border for trial divisions.

Shawe-Taylor's algorithm

1. If $2^k < L$, then to generate random odd k-bit number N and to test its primality with trial division. Repeat until a prime number N is generated. Exit.
2. If k is odd, let $k_1 = (k+3)/2$. If it is even, let $k_1 = k/2 + 1$. Recursive pass the algorithm with input parameter k_1 to construct a k_1-bit prime number.
3. Let x is random integer and $2^{k-1} \le x < 2^k$.
4. Let t is the smallest integer greater than $x/(2q)$.
5. If $2tq + 1 \ge 2^k$, then let t be the smallest integer greater than $2^{k-1}/(2q)$.
6. Let $N = 2tq + 1$.
7. Chose random integer a, $1 < a < N-1$, let $x = a^{2t} \bmod N$. If $x \ne 1$, $GCD(x-1, N) = 1$, $x^q = 1 \bmod N$, then N is generated prime.
8. Else let $t = t + 1$ and repeat steps 5–7.

However, the above algorithms guarantee only the primality of generated numbers. Additional requirements for numbers that are divisors of RSA module are the property of their so-called "maximization of the complexity of specialized factoring algorithms" (Zadiraka et al. 2007). For example, for $p \pm 1$ algorithms this property is revealed if n has prime factor p so that number $p \pm 1$ has sufficiently large power-smooth boundary (Mukhachev and Khoroshko 2005), in other words, $p \pm 1$ has large prime factors. This divisor p is called "strong" prime number. So we bring up an important question about the number and the distribution of such strong primes.

1.4 DISTRIBUTION AND AMOUNT OF STRONG PRIMES

Let p be a classic strong prime if the following requirements are met:

$$p = 1(\bmod r),\ p = -1(\bmod s),\ r = 1(\bmod t), \tag{1.1}$$

where r, s, t are large primes. This means that p, r, s, t can be shown as $p = 2jr + 1$, $p = 2ks - 1, r = 2lt + 1$ whereby the lower the numbers j, k, l the better.

Note that a generalization of the classical notion of prime number is a concept of strong primes that maximize the complexity of all known algorithms for factorization.

R.L. Rivest denied the idea of need to use strong primes as factors for RSA modules (Rivest and Silverman 1999). On the base of Lenstra's algorithm he wanted to show that the choice of p did not take an effect on the efficiency of obtaining the factorization, and that, in general, it is enough to choose random primes to obtain RSA modules. Later his ideas were refuted, since some algorithms, including ECM, work better if the number $p - 1$ is smooth.

Here follows the Gordon's method of generating strong prime numbers.

1. Construct a random prime number s of pre-selected size. We can select a pseudorandom number x of the desired size and using trial divisions we may leave the numbers in the range $[x, x+\log_2 x]$ that do not have small divisors. Among the remaining numbers choose prime number s with the help of primality test.
2. Generate random t in the same way.
3. Using trial division and primality test generate prime $r = 2lt+1$, sorting out l in $[1, \log_2 t]$.
4. Calculate $u = u(r,s) = (s^{r-1} - r^{s-1}) \bmod rs$.
5. If u is odd, then let $p_0 = u$, else $p_0 = u + rs$.
6. Test if $p = p_0 + 2krs$ is prime for $k = 0, 1, 2, \ldots$

Assume the condition of applying strong primes while generating cryptosystem's module. The following conditions for factors should be met:

$$\begin{matrix} p_1 \mid p-1, p_1 \geq n^{x_1}, p_2 \mid p_1 - 1, p_2 \geq n^{x_2}, p_3 \mid p+1, p_3 \geq n^{x_3}, 0 < x_1, x_2, x_3 < 0.5 \\ q_1 \mid q-1, q_1 \geq n^{x_4}, q_2 \mid q_1 - 1, q_2 \geq n^{x_5}, q_3 \mid q+1, q_3 \geq n^{x_6}, 0 < x_4, x_5, x_6 < 0.5 \end{matrix} \tag{1.2}$$

Let $p_i(n)$ be i-th the greatest divisor of n. Let $w_i(n,x) = \#\{t : 1 \leq t \leq n, p_i(t) \leq n^x\}$ is the amount of integers less or equal n for which $p_i(n)$ does not exceed n^x. Applying the result of Knuth and Trabb Pardo (1976) we have the probability that random number has the greatest divisor bigger than some border:

$$P\{q_1 > n^x\} = \int_x^1 \frac{dt}{t} = -\ln x. \tag{1.3}$$

Using Adamar and Vallee Poussin (Porter 1915) estimates we obtain the following estimate of the number of classic strong primes in $[a,b]$:

$$\pi_s(a,b) = -\ln x_1 \cdot \ln x_2 \cdot \ln x_3 \cdot \int_a^b \frac{dt}{\ln t}. \tag{1.4}$$

Considering the independence of the described divisors the estimate amount of RSA modules is defined as:

$$(\pi_s(a,b))^2. \tag{1.5}$$

Let us compare the resulting estimate of the number of RSA modules generated with random primes (no requirements for "strong" primes). RSA module is a so-called semiprime number because it has exactly two prime factors. The number whose prime factors less particular bound will be called B-smooth. Works (Ishmukhametov and Sharifullina 2014) present the distribution of semiprime and smooth numbers, the estimates are based on Riemann hypothesis. Probabilistic function $g(y)$ corresponds the probability that the number y will be "semiprime". Function approximation using series summation by prime numbers p:

$$g(y) \approx \sum_{p \leq \sqrt{y}} \frac{1}{p(\ln y - \ln p)}. \tag{1.6}$$

Using Mertens's formula:

$$\sum_{p<x} \frac{\ln p}{p} = \ln x + O(1), \tag{1.7}$$

and Abel's theorem results we have:

$$g(y) \approx \frac{\ln\ln y - \beta}{\ln y}, \quad \text{where } \beta = 1 + \ln\ln 2. \tag{1.8}$$

Let $\psi(x,y)$ be the amount of all y-smooth numbers less than x, we have the recurrent formula for the calculation:

$$\psi(x, p_k) \approx \sum_{i=0}^{t_k} \psi\left(\frac{x}{p_k^i}, p_{k-1}\right), t_k = \left[\frac{\ln x}{\ln p_k}\right], k > 1, \tag{1.9}$$

where p_k – k-th prime $(p_1 = 2), \psi(x,2) = \left[\frac{\ln x}{\ln 2}\right] + 1$.

There are rough estimates for calculating the number of smooth numbers because with the growth of smoothness bounds the computational complexity increases exponentially. Hildebrand (1986) obtained the following estimation:

$$\psi(x, y) \approx x \cdot \rho(u)\left\{1 + O\left(\frac{\ln(u+1)}{\ln y}\right)\right\}, \quad \text{where } x = y^u. \tag{1.10}$$

Here $\rho(u)$ is the Dickman-de Bruijn function, $u = \ln x / \ln y$, $\rho(u)$ is the solution of differential equation $u\rho'(u) + \rho(u-1) = 0$ for $u > 1$, this approximation is valid only with the simultaneous growth of x and y keeping $\ln x / \ln y$ as a constant. Here is the approximate formula for $y < (\ln x \ln\ln x)^{1/2}$:

$$\psi(x, y) \approx \frac{1}{\pi(y)!} \prod \left(\frac{\ln x}{\ln p}\right)\left\{1 + O\left(\frac{y^2}{\ln x \ln y}\right)\right\}. \tag{1.11}$$

But to estimate the capacity of the set of smooth numbers with practically used boundaries this approximation is inapplicable.

The considered estimations of semiprime number's distribution give an opportunity to present the size of the whole set. To solve the problem of RSA crypto parameters choice we should restrict this set only by choosing such semiprimes that are not smooth for some smoothness border B in order to maximize the complexity of solving the factorization problem of semiprime numbers.

In the above-mentioned notations we are interested in the following estimation for the number of semiprimes in the range $[2,T]$ that are not B-smooth (Ishmukhametov and Sharifullina 2014):

$$K(T, B) = \int_2^T \frac{\ln\ln x - \beta}{\ln x} dx - \psi(T, B), \quad \text{where } \beta = 1 + \ln\ln 2. \tag{1.12}$$

On the other hand it is possible to evaluate the amount of B-smooth numbers in $[2,T]$ with the help of combinatorial methods and group theory. It makes sense to choose primes as smoothness boundary. So having fixed $B = p_k$ as k-th prime number, the problem to evaluate the number of combinations $x = \prod_{i=1,k} p_i^{\alpha_i}$ where $x < T$. Note that the estimates (1.4) and (1.12) are close in value, but (1.4) is obtained in much easier way and takes into account the condition of choosing "strong" primes.

In general, the following statements are applicable for implementing RSA cryptosystems:

1. it is necessary to impose the restrictions for primes that are used for generation RSA modules for system's security in practice;
2. the output set algorithms that generate such primes have appreciably less cardinality of a set than cardinality of prime number set.

For these reasons, we formulate the hypothesis that there exists a polynomial complexity algorithm that enumerates the set of "secure" RSA modules. The above estimations support this hypothesis.

1.5 FEATURES OF FACTORIZATION ALGORITHM ELABORATION FOR CLOUD COMPUTING

The first problem of factorization price estimating for cloud computing is a form of ordering the services, which are typically given by providers. For example, Windows Azure provides services in a form of virtual machine with specific characteristics of the CPU, memory and disk space (Accessed June 3 2016. http://azure.microsoft.com/en-us/pricing/details/cloud-services). The most powerful architecture which is offered "for working applications with large databases, server applications and high-speed applications" (e.g., "D14" has the characteristics of 112 GB RAM, 800 GB of disk space, 16 core processor) costs $2.611 US per hour or about $1943 US per month.

The cost of computing module rent with random configuration for most cloud platforms may be estimated using the on-line calculators (Accessed June 3 2016. https://cloud.google.com/products/calculator, http://www.hpcloud.com/pricing). Thus, a computer module in the cloud Google Cloud Platform with the configuration similar to "D14" costs $5.36 US per hour, but the monthly rent will cost $689.08 (Accessed June 3 2016. http://azure.microsoft.com/en-us/pricing/details/cloud-services). In HP's Public cloud the configuration, aimed at calculating the maximum power (30 HP Cloud Compute Units, 4 virtual cores, 60GB RAM, 540GB ephemeral disk) is $1.35/hr ($985.50/mo.) (Accessed June 3 2016. http://www.hpcloud.com/pricing). You can also choose virtually unlimited (theoretically—only for the price) number of such machines.

So, to order a specific computing architecture targeted at particular factorization algorithm is difficult, so you need to rely on many computing nodes of standard minimum configuration aimed at powerful calculations that operate in parallel. The price of this node (e.g., n1-highcpu-2 for the cloud Google Cloud Platform) is $38.84 per month.

The second problem is that for factorization price estimating it is also necessary to choose the fastest at present, universal (which can be used for any kind of numbers) algorithm, which allows the simple implementation of a parallel computing model. The choice of this algorithm for factorization price estimating has similar value to the choice of the universal models of computation (Turing-Post machine, normal Markovian algorithm, brute-force program, etc.) for algorithms complexity estimate.

The generally accepted solution to this problem is the choice of algorithm GNFS (Buhler et al. 1993, Lenstra and Lenstra 1993, Couveignes 1993, Ishmukhametov and Sharifullina 2014). This algorithm, like the most effective factorization algorithms at present time operates quite a simple idea, proposed by Fermat, such as: if you find a pair of numbers (A,B), that satisfy the equation $A^2 - B^2 = n$, then $n = (A - B)(A + B)$. So, if $m = \lfloor \sqrt{n} \rfloor$, then x = 1,2,… calculate the value of polynomial $q(x) = (m + x)^2 - n$ until $q(x)$ is not a perfect square.

A simple generalization of Fermat's idea is to search for pairs of numbers that satisfy more general equation $A^2 - B^2 \equiv 0 \bmod (n)$ (Ishmukhametov and Sharifullina 2014). Along with this, it appears that the value of the polynomial $q(x)$ is the so-called "smooth" numbers. A pair of integers (A,B), is called "smooth" (on base factor F, formed of primes) if:

1. The relation $A^2 - B^2 \equiv 0 \bmod (n)$ is performed.
2. B is multiplied only by prime factors that belong to the set F.

Representation of polynomial values $q(x)$ by the elements of a vector space over F to search the full squares has led to Dixon algorithm first, then to quadratic sieve (QS), but as a generalization of the latter—the general number field sieve (GNFS). We will briefly present the main stages of this algorithm to discuss major evaluations of its performance (Ishmukhametov and Sharifullina 2014):

1. Choose irreducible polynomial degree $d \geq 3$.
2. We choose an integer $m, [n^{1/(d+1)}] < m < [n^{1/(d)}]$, and represent n on the basis of m:

$$n = m^d + a_{d-1}m^{d-1} + \ldots + a_0$$

3. With this representation we bind irreducible in the ring $Z[x]$ (the ring of polynomials of the variable x with integer coefficients) polynomial:

$$f_1(x) = x^d + a_{d-1}x^{d-1} + \ldots + a_0.$$

4. Define sieving polynomial $F_1(a,b)$ as a homogeneous polynomial of two variables a and b:

$$F_1(a,b) = b^d * f_1\left(\frac{a}{b}\right)$$

5. Also define another polynomial $f_2(x) = x - m$ and the corresponding homogeneous polynomial $F_2(a,b) = a - bm$. The main requirement for pair selection (f_1, f_2) is to fulfill the condition: $f_1(m) = f_2(m)\text{mod}(n)$ which is obviously performed in our case, because the first polynomial at the point m is n, and the second is zero.
6. Choose two positive numbers L_1 and L_2, which define the certain rectangular area $SR = \{1 \leq b \leq L_1, -L_2 \leq a \leq L_2\}$, called "sieve area".
7. Let θ – root of polynomial $f_1(x)$. Consider the polynomial ring $Z[\theta]$ (practically root θ is not evaluated, and is only used for formal description of the algorithm). Define the algebraic factor base FB_1, consisting of a first-order polynomial form $a - b\theta$ with the norm which is prime number. These polynomials are simple irreducibility elements in the ring of algebraic field $K = Q[\theta]$. The absolute value of the norms of polynomials from factor base FB_1 are bound above by some constant B_1.
8. At the same time let define rational factor base FB_2, consisting of all prime numbers bound above by second constant B_2.
9. To be able to check the final stage of the algorithm, whether the determined polynomial is perfect square, we define a relatively small set of polynomials $c - d\theta$ of the 1 order, the norm of which is also a simple number. Let this set be marked as FB_3. It must satisfy the condition $FB_1 \cap FB_3 = \varnothing$ is called Quadratic Character Factor Base.
10. Further simultaneous sieving of polynomials based on FB_1 is performed and integers based on FB_2 to obtain the set M, consisting of smooth pairs (a,b). The pair (a,b) is called smooth if $GCD(a,b) = 1$ and the polynomial $a - b\theta$ and the number $a - bm$ are multiplied entirely by the relevant factor bases FB_1 and FB_2. The number of smooth pairs in the set M must exceed the total power of three factor bases at least by two units.
11. The next step is to search the subset $S \subseteq M$ so that the product of all pairs $(a - b\theta)$ is H^2 and pairs $(a - bm)$ is B^2, $H \in Z, B \in Z$. To search the set S as in the quadratic sieve method, a system of linear algebraic equations is formulated with coefficients from of the set $F_2 = \{0,1\}$, the solution of which will be the numbers of the set S.
12. Next we form the polynomial:

$$g(\theta) = \left(f_1'(x)\right)^2 \prod\nolimits_{(a,b)\in S}(a - b\theta).$$

13. If the whole procedure is performed correctly, then the polynomial $g(\theta)$ is a perfect square in the ring of polynomials $Z[\theta]$. We extract square roots of the polynomial $g(\theta)$ and an integer B^2, finding some polynomial $\alpha(\theta)$ and the number B.
14. Replace the polynomial $\alpha(\theta)$ with the number $\alpha(m)$.
15. So, a pair of integers (A, B), satisfying the condition $A^2 - B^2 \equiv 0\,\text{mod}(n)$ has been found.

As seen from the description of the main stages of the algorithm there are two key parameters that affect its complexity—the size of the sieve area and the size of factor base. The complexity estimate of the algorithm GNFS is easy to measure by using $L_n(\alpha,c)=e^{\left(c+o(1)\left((\ln n)^{\alpha}(\ln\ln n)^{1-\alpha}\right)\right)}$ function. The most famous current estimates for the classical method GNFS – $L_n\left(\frac{1}{3};1{,}92..+o(1)\right)$ (Ishmukhametov and Sharifullina 2014), for the "group" GNFS (several numbers are factorized at a time for which common screening results, are used the average factorization time per each number is calculated) $L_n\left(\frac{1}{3};1{,}704+o(1)\right)$ (Bernstein and Lange 2014).

For approximate real cost estimate of factorization for the clouds given above let's give the estimate of the sieving area (the most complicated stage) of the classical method GNFS. It is $L_n\left(\frac{1}{3};\left(\frac{8}{9}\right)^{1/3}+o(1)\right)$.

If we take this estimate for the number of computing nodes of "minimal" configuration then the factorization cost estimate of 1024-bit number is hundred trillion US dollars, if the number is factorized during a month, which of course is much worse than the factorization cost estimates specially designed by using the devices for factorization (TWIRL or SHARK) (Shamir and Tromer 2003, Franke 2005). So the cost of the system SHARK (Franke 2005) for factorization of 1024-bit number is evaluated 160 million US dollars, if the number is factorized during a year.

Such estimate is explained first of all, by the lack of flexibility of cloud computing services order, and, secondly, by the direct use of existing implementations of the algorithm GNFS for cloud technologies. The doubtless advantage of cloud computing that determines the relevance of further research in this area, is their relatively greater availability compared to specialized calculators and opportunity to use the so-called "peak-mobilization" operation mode in "critical" cases—the use of cloud resources across the corporation, state and international association.

1.6 CONCLUSIONS

Existing standards of implementation and usage of asymmetric cryptosystems impose restrictions on the crypto parameters generation. This situation is conditioned by modern advances in solving the problem of factorization of large numbers. The interest of assessing RSA security in practice is the highlighted issue of size or rather the nature of size growth of the set of secure crypto parameters, namely semiprime numbers that have additional constraints about their divisors.

REFERENCES

Baydenko P.V., Kudin A.M., 2012: *Efektyvnist' zastosuvannya alhorytmiv faktoryzatsiyi do moduliv kryptosystemy RSA*, X Vseukrayins'ka naukovo-praktychna konferentsiya studentiv, aspirantiv ta molodykh vchenykh "Teoretychni i prykladni problemy fizyky, matematyky ta informatyky", Zbirka tez dopovidey uchasnykiv, Kyiv, Ukraine, pp. 253–254, (in Ukrainian).

Bernstein D.J., 2001: Circuits for integer factorization: a proposal, http://cr.yp.to/papers.html#nfscircuit.

Bernstein D.J., Chang Y.-A., Cheng C.-M., Chou L.-P., Heninger N., Lange T., van Someren N, 2013: *Factoring RSA keys from certified smart cards: Coppersmith in the wild*, Cryptology ePrint Archive, Report 2013/599, https://eprint.iacr.org/2013/598.pdf.

Bernstein D.J., Lange T., 2014: *Batch NFS*, Selected Areas in Cryptography, LNCS, Springer, vol. 8781, pp. 38–58, http://cr.yp.to/factorization/batchnfs-20141109.pdf.

Brent R.P., 1990: Parallel algorithms for integer factorization, Number Theory and Cryptography (edited by J.H. Loxton), London Mathematical Society Lecture Notes (Book 154), Cambridge University Press, pp. 26–37.

Buhler J.P., Hendrik W., Lenstra Jr., Pomerance C., 1993: *Factoring integers with the number field sieve*, The Development of the Number Field Sieve, Springer, vol. 1554, pp. 50–94.
Couveignes J., 1993: *A general number sieve implementation*, Lecture Notes in Math., vol. 1554, pp. 103–126.
Crandall R.E., 1999: Parallelization of Pollard-rho factorization, http://academic.reed.edu/physics/faculty/crandall/papers/parrho.pdf.
Digital signature standard, National Institute of Standards and Technology, Federal Information Processing Standards Publication, FIPS PUB 186-4-1.
Franke J., Kleinjung T., Paar C., Pelzl J., Priplata C., Stahlke C., 2005: *SHARK: a realizable special hardware sieving device for factoring 1024-bit integers*, Cryptographic hardware and embedded systems, LNCS, Springer, vol. 3659, pp. 119–130.
Google Cloud Platform pricing calculator, https://cloud.google.com/products/calculator.
Gordon J., 1984: *Strong RSA keys*, Electronics letters, vol. 20(12), pp. 514–516.
Hartman W.J., 1982: *An Experimental Study of Monte Carlo Factoring Techniques*, National Telecommunications and Information Administration.
Heninger N., Durumeric Z., Wustrow E., Halderman J.A., 2012: *Mining Your Ps and Qs: detection of widespread weak keys in network devices*, In Proceedings of the 21th USENIX Security Symposium, Bellevue, WA, USA, August 8–10, pp. 205–220.
Hildebrand A., 1986: *On the number of positive integers* $\leq x$ *and free of prime factors* $> y$, Journal of Number Theory, vol. 22(3), pp. 289–307.
HP Cloud pricing, http://www.hpcloud.com/pricing.
Ishmukhametov Sh.T., Sharifullina F.F., 2014: *On distribution of semiprime numbers*, Russian Mathematics, vol. 58(8), pp. 43–48.
Knuth, Donald E., et al.: Selected Papers on Computer Languages, Stanford, CA, CSLI, 2003. ISBN 1-57586-382-0) (typewritten draft, August 1976).
Koundinya A.K., Harish G., Srinath N.K., Raghavendra G.E., Pramod Y.V., Sandeep R., Kumar P.G., 2013: *Performance analysis of parallel Pollard's RHO factoring algorithm*, International Journal of Computer Science & Information Technology (IJCSIT), vol. 5(2), pp. 157–164.
Lenstra A.K., Lenstra H.W., 1993: *The development of Number Field Sieve*, LNM, Springer-Verlag, vol. 1554.
Lenstra A.K., Hughes J.P., Augier M., Bos J.W., Kleinjung T., Wachter C., 2012: Ron was wrong, Whit is right, Cryptology ePrint Archive, Report 2012/064, http://eprint.iacr.org/2012/064.pdf.
Lenstra Jr. H.W., 1987: *Factoring integers with elliptic curves*, Annals of Mathematics, Second Series, vol. 126(3), pp. 649–673.
Maurer U.M., 1990: *Fast generation of secure RSA-moduli with almost maximal diversity*, Advances in Cryptology—EUROCRYPT '89, LNCS, Spinger-Verlag, vol. 434, pp. 636–647.
Microsoft Azure. Cloud services pricing, http://azure.microsoft.com/en-us/pricing/details/cloud-services.
Mukhachev V.A., Khoroshko V.A., 2005: *Metody prakticheskoy kriptografii*, Poligraf-Konsalting, Kyiv, Ukraine, (in Ukrainian).
Pollard J.M., 1975: *A Monte Carlo method for factorization*, BIT Numerical Mathematics, vol. 15(3), pp. 331–334.
Porter, M.B. 1915. Review: Cours d'Analyse Infinitésmale, by Ch.-J. de la Vallée Poussin. Bull. Amer. Math. Soc. 22. pp. 77–85.
Rivest R.L., Silverman R.D., 1999: *Are "strong" primes needed for RSA?* RSA Laboratories Seminar Series, Seminars Proceedings.
Sekey G., 1993: Paradoksy v teorii veroyatnostey i matematicheskoy statistike, Mir, Moscow, Russia (in Russian).
Shamir A., 1999: *Factoring large numbers with the TWINKLE device*, Cryptographic Hardware and Embedded Systems, LNCS, Springer-Verlag, vol. 1717, pp. 2–12.
Shamir A., Tromer E., 2003: *Factoring large numbers with the TWIRL device*, Advances in Cryptology—CRYPTO 2003, LNCS, Springer-Verlag, vol. 2729, pp. 1–26.
Shawe-Taylor J., 1986: *Generating strong primes*, Electronic letters, vol. 22(16), pp. 875–877.
Shor P.W., 1999: *Polynomial Time Algorithms for Prime Factorization and Discrete Logarithms on a Quantum Computer*, SIAM Review, vol. 41(2), pp. 303–332.
Zadiraka V.K., Kudin A.M., Oleksyuk A.S., 2007: *Adaptivnye algoritmy polucheniya prostykh chisel i ikh primenenie v kriptografii*, Komp'yuternaya matematika, no. 1, pp. 54–61, (in Ukrainian).
Zadiraka V.K., Kudin A.M., 2013: *Cloud computing in cryptography and steganography*, Cybernetics and Systems Analysis, vol. 49(4), pp. 584–588.

CHAPTER 2

Game problems of control for functional-differential systems

A.A. Chikrii, G.Ts. Chikrii & V.J. Zhukovskij
V.M. Glushkov Institute of Cybernetic of NAS of Ukraine, Kyiv, Ukraine

W. Wójcik
Lublin University of Technology, Lublin, Poland

M. Junisbekov
M.Kh. Dulaty Taraz State University, Taraz, Kazakhstan

INTRODUCTION

The strategy of creating solution (control choice) based on information about the current position of the dynamic process goes back to the classic works of N.N. Krasovskii. These are his monograph (Krasovskii 1970) and the subsequent book in coauthorship with A.I. Subbotin (Krasovskii and Subbotin 1974). Employment of the Extremal Targetting Rule, developed in (Krasovskii 1970), allows for the game termination in the "first absorption" time in the regular and regularized cases. Similar result was obtained by B.N. Pshenichnyi (1968) with the use of the convex analysis technique. On the one hand, this method is the result of extension of the Pontryagin Maximum Principle to the game problems. On the other hand, it justifies the Euler law of goal interception, known as the "Line of Sight Law".

The Extremal Targetting Rule appeared to be very fruitful. Many scientists did research in this direction. B.N. Pshenichnyi Yu.N. Onopchuk adopted this method to the case of integral constraints on controls (Pshenichnyi and Onopchuk 1968). G.Ts. Chikrii extended the result of B.N. Pshenichnyi (1968) to the non-stationary case (Chikrii G.Ts. 1995). J.S. Rappoport studied the case of mixed controls (Krasovskii and Subbotin 1974). Positional conflict control by the systems of integral and integro-differential equations was studied in the papers of V.L. Pasikov (Krasovskii and Subbotin 1974), G.Ts. Chikrii and K. Volyanskij (2002). Complex positional problems of the group pursuit were first stated and solved in the works of S.I. Tarlinskij (1976), A.A. Chikrii (1979), A.A. Chikrii and J.S. Rappoport (1978). A.Ye. Perekatov and A.A. Chikrii (1993) examined the problem of positional pursuit of a group of evaders by controlled moving object (the problem of "commercial traveler" type).

Investigations of M.S. Gabrielyan and A.V. Kryazhimskii (1983) are devoted to the study of positional conflict counteraction of controlled objects groups. It should be noted that, from mathematical point of view, the Extremal Targetting Rule, as applied to linear systems, is based on using the apparatus of support functions, the notion of Aumann integral of set-valued mapping and the Lyapunov theorem on vector measures.

Due to the positional formalization, N.N. Krasovskij (1985) proved the theorems of alternative, dividing the game space into the players' preference domains. In the case of failure of the "saddle point condition in small game" the alternative can be formulated in terms of strategy of one player and counterstrategy of another. Under the same condition, an assertion on alternative can be stated in mixed strategies.

Because of the problem on existence of solution of the differential equations, arising in such formalization, B.N. Pshenichnyi (1969) introduced the notion of ε-strategy being the development of ideas of Pontryagin's (1988) Alternated Integral Method. Thus, there exist many methods of constructing positional controls in conflict condition.

In this chapter, we suggest a general scheme for investigation of conflict-controlled processes, illustrated on various types of functional-differential systems.

Note that application of the method of resolving functions (A.A. Chikrii 1997) to similar problems is based on another assumption on information availability, that leads to control choice in the class of counter- or quasi-strategies.

2.1 PROBLEM STATEMENT, TIME OF THE "FIRST ABSORPTION"

Let a state of the conflict-controlled process at arbitrary moment of time t, $t \geq t_0 \geq 0$, be described by the vector:

$$z(t) = g(t) + \int_{t_0}^{t} \phi(t,s,u(s),v(s))ds, \tag{2.1}$$

where $z(t) \in R^n, g(t), g:[t_0,+\infty) \to R^n$, is continuous vector-function, $\phi(t,s,u,v), \phi: \Delta(t_0) \times U \times V \to R^n$, – function, jointly continuous in its variables. Here:

$$\Delta(t_0) = \{(t,s): t_0 \leq s \leq t < +\infty\}, U \in K(R^n), V \in K(R^n).$$

By admissible controls of the first and second players are meant measurable functions, taking its values in U, V, respectively. Let us introduce the notations:

$$\Omega_U = \{u(s): u(s) \in U, s \in [t_0,+\infty)\}, \Omega_V = \{v(s): v(s) \in V, s \in [t_0,+\infty)\},$$

where $u(s)$ and $v(s)$ are measurable functions. We denote by $u(\cdot)$ and $v(\cdot)$ elements of the sets Ω_U, Ω_V, respectively.

Also, a cylindrical terminal set M^*, $M^* \subset R^n$, is specified:

$$M^* = M_0 + M, \tag{2.2}$$

where M^0 is a linear subspace in R^n and M is a convex compact from the orthogonal complement L to M^0 in R^n, $M \in coK(L)$. We suppose that $g(t_0) \notin M^*$.

The goal of the pursuer (u) is to bring a trajectory of the object (2.1) to the terminal set in the shortest time, whereas the evader (v) tries to avoid the encounter of the trajectory with the set. The game is terminated at the moment T, $T > t_0$, such that $z(T) \in M^*$ or $\pi z(T) \in M$.

Let us take the side of the first pursuer side and find the guaranteed time of the game termination. We assume that on the half-interval $[t_0, t)$ and at $t \geq t_0$ the players employ admissible controls:

$$u_t(\cdot) = \{u(s): s \in [t_0,t)\}, v_t(\cdot) = \{v(s): s \in [t_0,t)\}, u(\cdot) \in \Omega_U, v(\cdot) \in \Omega_V.$$

Then, for arbitrary finite moments t and T, $t_0 \leq t \leq T$, and fixed controls $u_t(\cdot)$ and $v_t(\cdot)$ following representation is valid:

$$z(T) = z(T,t,u_t(\cdot),v_t(\cdot)) + \int_{t}^{T} \phi(T,s,u(s),v(s))ds, \tag{2.3}$$

where:

$$z(T,t,u_t(\cdot),v_t(\cdot)) = g(T) + \int_{t_0}^{t} \phi(T,s,u(s),v(s))\, ds. \tag{2.4}$$

Note that $z\left(T,t_0,u_{t_0}(\cdot),v_{t_0}(\cdot)\right)=g(T)$.
Let us take into consideration a scalar product:

$$\left(z(T),p\right)=\left(z\left(T,t,u_t(\cdot),v_t(\cdot)\right),p\right)+\int_t^T\left(\phi\left(T,s,u(s),v(s)\right),p\right)ds,\quad p\in R^n.$$

Now we proceed to analysis of the following expression:

$$\begin{aligned}&\min_{v(\cdot)\in\Omega_V}\max_{u(\cdot)\in\Omega_U}\left(z(T),p\right)\\&=\left(z\left(T,t,u_t(\cdot),v_t(\cdot)\right),p\right)+\min_{v(\cdot)\in\Omega_V}\max_{u(\cdot)\in\Omega_U}\int_t^T\left(\phi\left(T,s,u(s),v(s)\right),p\right)ds\\&=\left(z\left(T,t,u_t(\cdot),v_t(\cdot)\right),p\right)+\int_t^T\min_{v\in V}\max_{u\in U}\left(\phi(T,s,u,v),p\right)ds.\end{aligned}$$

The signs of extremum and integral in the expression above can be interchanged in view of the properties of function $\phi(T, s, u, v)$ and the Lyapunov theorem on vector measures (Ioffe and Tikhomirov 1979). Denote:

$$\zeta=z\left(T,t,u_t(\cdot),v_t(\cdot)\right),$$

and consider the following function:

$$W(T,t,\zeta;p)=(\zeta,p)+\int_t^T\min_{v\in V}\max_{u\in U}\left(\phi(T,s,u,v),p\right)ds,\ t_0\le t\le T,\ \zeta\in R^n,\ p\in R^n.$$

It is obvious that function $W(T, t, \zeta, p)$ is continuous in the domain of its definition. Its vector-gradient in t, ζ has the form:

$$\nabla_{t,\zeta}W(T,t,\zeta;p)=\begin{pmatrix}\nabla_t W(T,t,\zeta;p)\\ \nabla_\zeta W(T,t,\zeta;p)\end{pmatrix}=\begin{pmatrix}-\min_{v\in V}\max_{u\in U}\left(p,\phi(T,t,u,v)\right)\\ p\end{pmatrix}.$$

Function $W(T, t, \zeta, p)$ is continuous in T, t, p by joint continuity of the function $\phi(t, s, u, v)$. Consider the support function of cylindrical set M^*.

$$C(M^*;p)=\begin{cases}C(M;p), & p\in L\\ +\infty & ,p\notin L.\end{cases}$$

The subspace L is the barrier cone of the set M. Because M is a compact lying in L, the function $C(M, p)$ is continuous on L.
Now we introduce a function:

$$\lambda(T,t,\zeta)=\min_{\|p\|=1,p\in L}\left[W(T,t,\zeta;p)+C(M;-p)\right]\qquad(2.5)$$

Using the notations:

$$T\left(t,u_t(\cdot),v_t(\cdot)\right)=\min\left\{T\ge t:\lambda\left(T,t,z\left(T,t,u_t(\cdot),v_t(\cdot)\right)\right)=0\right\},\qquad(2.6)$$

$$\Gamma(T,t,\zeta)=\{p: p\in L,\|p\|=1,\quad W(T,t,\zeta;p)+C(M;-p)=\lambda(T,t,\zeta)\},\tag{2.7}$$

let us introduce a set-valued mapping:

$$W(T,t)=\bigcap_{v(\cdot)\in\Omega_V}\left(M-\int_t^T\pi\,\phi(T,s,U,v(s))\,ds\right).\tag{2.8}$$

The above expression contains integral of a set-valued mapping, namely, the Aumann integral (Aumann 1965). By virtue of the properties of Aumann integral (Aubin and Frankowska 1990), the set-valued mapping $W(T, t)$ is convex-valued and upper semi-continuous.

The so-called time of the "first absorption" is defined in the following way:

$$T_0=\min\{T: T\geq t_0, \pi\, g(T)\in W(T,t_0)\}$$

We assume that the minimum is attained.

Let us evaluate the support function of the set $W(T, t_0)$:

$$C(W(T,t_0);p)=co\left[C(M;p)+\int_{t_0}^T\min_{v\in V}\max_{u\in U}(-p,\phi(T,s,u,v))ds\right].$$

It is evident that $g(t)\in W(T,t_0)$ if and only if:

$$\min_{p\in L,\|p\|=1}\left[C(W(T,t_0);-p)+(p,g(T))\right]\geq 0.$$

As the minimum of function coincides with the minimum of its convexification we have:

$$\min_{p\in L,\|p\|=1}\left[C(W(T,t_0);-p)+(p,g(T))\right]=\lambda(T,t_0,g(T)).$$

This means that:

$$T_0=T\left(t_0,u_{t_0}(\cdot),v_{t_0}(\cdot)\right)=\min\{T\geq t_0:\lambda(T,t_0,g(T))=0\}.$$

Let us suppose that the minimum is attained. We assume that the time of "first absorption" is finite, that is $T_0<+\infty$. That the time T_0 is the minimal instant of time at which $\pi\, g\,(T_0)\in W\,(T_0,t_0)$, means that T_0 is the minimal time in which the first player, knowing beforehand the second player's control for all the time ahead, can steer the system trajectory to the terminal set M^*. Let us show that $T_0>t_0$. Using formulas (2.5), (2.8), we obtain:

$$\lambda\left(t_0,t_0,z\left(t_0,t_0,u_{t_0}(\cdot),v_{t_0}(\cdot)\right)\right)=\min_{\|p\|=1,p\in L}\left[W\left(t_0,t_0,g(t_0),p\right)+C(M;-p)\right]$$

As $g(t_0)\notin M$, then $\lambda(t_0,t_0,g(t_0))<0$, whence follows that $T_0\neq t_0$.

Lemma 2.1. *If* $T(t,u_t(.),v_t(.))=t$, *then* $z(t)\in M$.

Proof. Let on the half-interval $[t_0,t)$ the players employed the controls $u_t(\cdot)$, $u_t(\cdot)\in\Omega_U$ and $v_t(\cdot)$, $v_t(\cdot)\in\Omega_V$. If $T(t,u_t(\cdot),v_t(\cdot))=t$, then there follows from the definition (2.5) that $\lambda(t,t,z(t,t,u_t(\cdot),v_t(\cdot)))\geq 0$. By the formula (2.4):

$$z\left(t,t,u_t(\cdot),v_t(\cdot)\right)=g(t)+\int_{t_0}^t\phi(t,s,u(s),v(s))ds,$$

therefore, $\{\pi z(t)\}\cap M\neq\varnothing$ and $z(t)\in M$.

2.2 SUFFICIENT CONDITIONS FOR THE GAME TERMINATION

Let us introduce the notation:

$$T_t = T\left(t, u_t(\cdot), v_t(\cdot)\right).$$

We call the pair:

$$\left(t, z\left(T_t, t, u_t(\cdot), v_t(\cdot)\right)\right) \in R^{n+1}, t \geq t_0,$$

a current position of the game. Then an initial position is the pair:

$$\left(t_0, z\left(T_0, t_0, u_{t_0}(\cdot), v_{t_0}(\cdot)\right)\right) = \left(t_0, g(T_0)\right).$$

There follows from the time of "first absorption" that the first player is aware of the second player control for all the time ahead. Below conditions are established, insuring that the first player can terminate the game no later than now of "first absorption", under the condition that the first player constructs its control based on information on current position of the game.

Theorem 2.1. *Let for the game* (2.1), (2.2) *following conditions hold:*

1. *set-valued mapping $W(T, t)$ is non-empty ($W(T, t) \neq \varnothing$), the time of "first absorption" is finite ($T_0 < +\infty$), $t_0 \leq t \leq T \leq T_0$;*
2. *for any position $\left(t^*, z\left(T_{t^*}, t^*, u_{t^*}(\cdot), v_{t^*}(\cdot)\right)\right)$, $t_0 \leq t^* \leq T_0$, such that $T_{t^*} \leq T_0$, a neighborhood $\Omega\left(t^*, z\left(T_{t^*}, t^*, u_{t^*}(\cdot), v_{t^*}(\cdot)\right)\right)$ of the point $\left(t^*, z\left(T_{t^*}, t^*, u_{t^*}(\cdot), v_{t^*}(\cdot)\right)\right)$ exists, such that for all $(t, \zeta) \in \Omega\left(t^*, z\left(T_{t^*}, t^*, u_{t^*}(\cdot), v_{t^*}(\cdot)\right)\right)$ the set $\Gamma(T_{t^*}, t, \zeta)$* (2.7) *consists of a unique element $p(T_{t^*}, t^*, \zeta)$.*

Then the first player can terminate the game from given initial position no later than at the time T_0 under arbitrary admissible control of the second player.

Proof. First consider the case $t_* = t_0$. For any fixed $T \geq t_0$ the function $z\left(T, t, u_t(\cdot), v_t(\cdot)\right)$, $t_0 \leq t \leq T$, is continuous in t. In view of assumption 2) of the theorem there exists a moment of time $t_1, t_1 > t_0$, such that for all possible $z\left(T_0, t, u_t(\cdot), v_t(\cdot)\right), t \in [t_0, t_1)$, the set $\Gamma\left(T_0, t, z\left(T_0, t, u_t(\cdot), v_t(\cdot)\right)\right)$ (2.7) consists of a unique element $p_0\left(T_0, t, z\left(T_0, t, u_t(\cdot), v_t(\cdot)\right)\right)$. Now we will analyze the set-valued mapping:

$$\begin{aligned} U^e\left(T_0, p_0\left(T_0, t, \zeta\right), t, v(t)\right) = \Big\{ u : u \in U, \ & \max_{u \in U}\left(p_0\left(T_0, t, \zeta\right), \phi\left(T_0, t, u, v(t)\right)\right)\Big\} \\ = \left(p_0\left(T_0, t, \zeta\right), \phi\left(T_0, t, u, v(t)\right)\right), \ & t \in [t_0, t_1). \end{aligned} \tag{2.9}$$

Above it was used the notation:

$$\zeta = z\left(T_0, t, u_t(\cdot), v_t(\cdot)\right).$$

The set-valued mapping (2.9) is compact-valued and upper semi-continuous by inclusion in variables (p,t,v) by assumptions concerning parameters of the conflict-controlled process (2.1) (Aubin and Frankowska 1990). That is why, by the theorem on measurable choice the set-valued mapping $U^e\left(T_0, p_0, t, v\right)$ contains at least one Borel selection $u^e\left(p_0, t, v\right)$.

Since $z\left(T_0, t, u_t(\cdot), v_t(\cdot)\right)$ is continuous in t, vector-function $p_0\left(T_0, t, z\left(T_0, t, u_t(\cdot), v_t(\cdot)\right)\right)$ is continuous in $t, t \in [t_0, t_1)$ as well. From the control $v(t), t \in [t_0, t_1)$, measurability and the vector $p_0\left(T_0, t, z\left(T_0, t, u_t(\cdot), v_t(\cdot)\right)\right)$ continuity in t there follows that the Borel selection $u^e\left(p_0, t, v\right)$ is superpositionally measurable function. Consequently, function:

$$u^e(t) = u^e\left(p_0\left(T_0,t,z\left(T_0,t,u_t(\cdot),v_t(\cdot)\right)\right),v(t)\right), t \in [t_0,t_1)$$

is measurable. We choose the pursuer control on the half-interval $[t_0,t_1)$ in the form of selection $u^e(t), t \in [t_0,t_1)$. In the case there exist several selections, an arbitrary one can be used.

According to the properties of minimum, for each $T, T \geq t_0$, the function $\lambda(T,t,\zeta)$, $t_0 \leq t \leq T, \zeta \in R^n$, is differentiable in t,ζ along any direction. By the second assumption the set $\Gamma(T,t,\zeta)$ consists of a unique element $p(T,t,\zeta)$ for all (t,ζ) from some neighborhood $O(t^*,\zeta^*)$ of the point (t^*,ζ^*). From this it follows that the vector-function $p(T,t,\zeta)$ continuously depends on (t,ζ) and:

$$\partial_{t,\zeta}\lambda(T,t,\zeta) = \begin{pmatrix} -\min\limits_{v\in V}\max\limits_{u\in U}\left(p(T,t,\zeta),\phi(T,t,u,v)\right) \\ p(T,t,\zeta) \end{pmatrix} \quad \forall(t,\zeta) \in O(t^*,\zeta^*).$$

Now we show that, under the control $u^e(t), t \in [t_0,t_1)$, and arbitrary admissible control $v(t), t \in [t_0,t_1)$, the system (2.1) at time t_1 come to the position:

$$\left(t_1, z\left(T_{t_1},t_1,u_{t_1}(\cdot),v_{t_1}(\cdot)\right)\right),$$

and $T_0 \geq T_{t_1}$. To this end, we evaluate the derivative:

$$\frac{d}{dt}\lambda\left(T_0,t,z\left(T_0,t,u_t^e(\cdot),v_t(\cdot)\right)\right), t \in [t_0,t_1).$$

To simplify calculations, we denote:

$$z\left(T_0,t,u_t^e(\cdot),v_t(\cdot)\right) = \zeta^e.$$

By the formula (2.4) and the fact that for any fixed $T, T \geq t_0$, the function $z\left(T,t,u_t(\cdot),v_t(\cdot)\right)$ is differentiable in t on the half-interval $[t_0,t_1)$ with:

$$\frac{d}{dt}z\left(T,t,u_t(\cdot),v_t(\cdot)\right) = \phi\left(T,t,u(t),v(t)\right),$$

we come to conclusion that:

$$\begin{aligned}\frac{d}{dt}\lambda(T_0,t,\zeta^e) &= -\min_{v\in V}\max_{u\in U}\left(p_0(T_0,t,\zeta^e),\phi(T_0,t,u(t),v(t))\right) + \left(p_0(T_0,t,\zeta^e),\phi(T_0,t,u^e(t),v(t))\right) \\ &= -\min_{v\in V}\left(p_0(T_0,t,\zeta^e),\phi(T_0,t,u^e(t),v(t))\right) + \left(p_0(T_0,t,\zeta^e),\phi(T_0,t,u^e(t),v(t))\right) \geq 0.\end{aligned}$$

In taking derivative of the function $\lambda(T_0,t,\zeta^e)$ the assumption that:

$$\Gamma\left(T_{t^*},t,z\left(T_0,t,u_t(\cdot),v_t(\cdot)\right)\right),$$

at each moment contains a unique element plays a crucial role. Thus,

$$\frac{d}{dt}\lambda\left(T_0,t,z\left(T_0,t,u_t^e(\cdot),v_t(\cdot)\right)\right) \geq 0, \quad t \in [t_0,t_1).$$

Therefore, function $\lambda\left(T_0,t,z\left(T_0,t,u_t^e(\cdot),v_t(\cdot)\right)\right)$ does not decrease on the half-interval $[t_0,t_1)$. Now we show that under the control $u^e(t), t \in [t_0,t_1)$, of the first player and arbitrary admissible control $v(t), t \in [t_0, t_1)$, of the second player, the game arrives at the position $(t_1, z(T_{t_1}, t_1, u_{t_1}(\cdot), v_{t_1}(\cdot)))$, for which $T_0 \geq T_{t_1}$. Since $\lambda(T_0, t_0, z(T_0, t_0, u^e_{t_0}(\cdot), v_{t_0}(\cdot))) \geq 0$ then:

$$\lambda(T_0, t_1, z(T_0, t_1, u^e_{t_1}(\cdot), v_{t_1}(\cdot))) \geq 0$$

The latter inequality indicates that according to (2.6):

$$T_{t_1} \leq T_0, \quad T(t_1, u^e_{t_1}(\cdot), v_{t_1}(\cdot)) \leq T_0.$$

It is not improbable that the equality $T(t, u_t(\cdot), v_t(\cdot)) = t$ is fulfilled on the half-interval $[t_0, t_1)$ that would mean termination of the game (lemma 2.1). Otherwise, by assumptions of the theorem 2.1, there exists $t_2, t_2 > t_1$, such that on the half-interval $[t_1, t_2)$ a replica of finding a selection of the set-valued mapping:

$$U^e(T_{t_1}, p_0(T_{t_1}, t, \zeta), t, v(t)) = \Big\{ u : u \in U, \max_{u \in U}(p_0(T_{t_1}, t, \zeta), \phi(T_{t_1}, t, u, v(t)))$$
$$= (p_0(T_{t_1}, t, \zeta), \phi(T_{t_1}, t, u, v(t)))\Big\}, t \in [t_1, t_2), \zeta = \breve{z}(T_{t_1}, t, u_t(\cdot), v_t(\cdot)),$$

can be made. The possibility exists that on the half-interval $[t_1, t_2)$ the equality $T(t, u^e_t(\cdot), v_t(\cdot)) = t$ occurs. In such case the game is terminated.

Otherwise, an instant t_3, $t_3 > t_2$, exists such that the procedure of constructing control u can be repeated. Keeping this process in progress, we obtain a consequence of the half-intervals, control of the pursuer on each of them is chosen in the form of selection $u^e(t)$ of the set-valued mapping:

$$U^e(T_{t_k}, p_0(T_{t_k}, t, \zeta), t, v(t)) = \Big\{ u : u \in U, \max_{u \in U}(p_0(T_{t_k}, t, \zeta), \phi(T_{t_k}, t, u, v(t)))$$
$$= (p_0(T_{t_k}, t, \zeta), \phi(T_{t_k}, t, u, v(t)))\Big\}, \zeta = z(T_{t_k}, t, u_t(\cdot), v_t(\cdot))\, t \in [t_k, t_{k+1}), t_{k+1} \leq T_{t_k}.$$

Since the set-valued mapping $U^e(T, p_0, t, v)$ is compact-valued and is upper semi-continuous in (p_0, t, v) for any T, it contains a Borelean selection $u^e(p_0, t, v)$, which is a superpositionally measurable function. Taking into account the fact that $p_0(T_{t_k}, t, \zeta)$ is continuous in (t, ζ) and $\zeta = \zeta(t)$ is continuous in t, we conclude that $p_0(T_{t_k}, t, \zeta(t))$ is continuous in t and, therefore, $u^e(t)$ appears as a measurable function on each of the half-intervals $[t_k, t_{k+1})$.

According to the results of (Natanson 1957) the function $u^e(t)$ is measurable on each of the half-intervals $t \in [t_k, t_{k+1})$, $k = 0, 1, 2, \ldots$, therefore, it is measurable on their countable union, namely, at $t \geq t_0$. Let us show that the following relationships are true:

$$T(t_k, u^e_{t_k}(\cdot), v_{t_k}(\cdot)) \geq T(t_{k+1}, u^e_{t_{k+1}}(\cdot), v_{t_{k+1}}(\cdot)). \tag{2.11}$$

To this end, we evaluate:

$$\frac{d}{dt}\lambda(T_{t_k}, t, z(T_{t_k}, t, u^e_{t_k}(\cdot), v_{t_k}(\cdot))), \; t \in [t_k, t_{k+1}).$$

Let us denote:

$$z(T_{t_k}, t, u^e_{t_k}(\cdot), v_{t_k}(\cdot)) = \zeta^e.$$

The procedure analogous to that performed on the half-interval $[t_0, t_1)$, yields:

$$\frac{d}{dt}\lambda\left(T_{t_k},t,\zeta^e\right)=-\min_{v\in V}\max_{u\in U}\left(p_0\left(T_{t_k},t,\zeta^e\right),\phi\left(T_{t_k},t,u,v\right)\right)+\left(p_0\left(T_{t_k},t,\zeta^e\right),\phi\left(T_{t_k},t,u^e(t),v(t)\right)\right)$$
$$=-\min_{v\in V}\left(p_0\left(T_{t_k},t,\zeta^e\right),\phi\left(T_{t_k},t,u^e(t),v\right)\right)+\left(p_0\left(T_{t_k},t,\zeta^e\right),\phi\left(T_{t_k},t,u^e(t),v(t)\right)\right)\geq 0.$$

Thus,

$$\frac{d}{dt}\lambda\left(T_{t_k},t,z\left(T_{t_k},t,u_t^e(\cdot),v_t(\cdot)\right)\right)\geq 0,\quad t\in\left[t_k,t_{k+1}\right),$$

function $\lambda\left(T_{t_k},t,z\left(T_{t_k},t,u_t^e(\cdot),v_t(\cdot)\right)\right)$ do not decrease on the on the half-interval $\left[t_k,t_{k+1}\right)$. By definition, $\lambda\left(T_{t_k},t_k,z\left(T_{t_k},t,u_{t_k}^e(\cdot),v_{t_k}(\cdot)\right)\right)\geq 0$, that is why:

$$\lambda\left(T_{t_k},t_{k+1},z\left(T_{t_k},t_{k+1},u_{t_{k+1}}^e(\cdot),v_{t_{k+1}}(\cdot)\right)\right)\geq 0.$$

Here from there follows inequality (2.11) that means:

$$T\left(t,u_t^e(\cdot),v_t(\cdot)\right)\leq T_0,\ t_0\leq t\leq T_0.$$

On the other hand, according to (2.6), $T\left(t,u_t^e(\cdot),v_t(\cdot)\right)\geq t,\ t_0\leq t\leq T_0$.
Thus, we have:

$$t\leq T\left(t,u_t^e(\cdot),v_t(\cdot)\right)\leq T_0,\quad t_0\leq t\leq T_0. \tag{2.12}$$

Therefore, the equality:

$$T\left(t,u_t^e(\cdot),v_t(\cdot)\right)=t, \tag{2.13}$$

indicative of the game termination (lemma 2.1), is satisfied no later than at the time T_0.

Suppose that this procedure lasts infinitely long. Then we obtain a consequence of half-intervals $\left[t_0,t_1\right),\left[t_1,t_2\right),\left[t_2,t_3\right),\ldots$. It is evident that $t_k<T_0,\ k=0,1,2,\ldots$, otherwise the equality (2.13) would be satisfied on one of the half-intervals and the game would be terminated. The $\{t_k\},\ k=0,1,2,\ldots$, is monotonically increasing and bounded above sequence that is why there exists a moment of time $\overline{t}$ such that $\lim_{k\to+\infty}t_k=\overline{t}$. It is clear that $T\left(\overline{t},u_{\overline{t}}^e(\cdot),v_{\overline{t}}(\cdot)\right)\leq T_0$. The procedure of constructing control u can be extended on some half-interval $\left[\overline{t},\overline{\overline{t}}\right),\ \overline{\overline{t}}>\overline{t}$, beginning from the position $\left(\overline{t},z\left(T_{\overline{t}},\overline{t},u_{\overline{t}}^e(\cdot),v_{\overline{t}}(\cdot)\right)\right)$.

Note that various controls $u_t^e(\cdot),t\in\left[t_k,t_{k+1}\right)$, chosen, which are measurable selections of the set-valued mapping (2.10), generate different system trajectories. Evidently, the double inequality (2.12) holds for all motions, induced by the controls $u_t^e(\cdot),t\in\left[t_k,t_{k+1}\right),\ k=0,1,2,\ldots$. In view of this inequality the condition (2.13), indicative to the game termination, will be fulfilled no later than at the time T_0. The theorem is proved.

Remark 2.1. *Let in the game under study an analog of the "small game" saddle point condition (Krasovskii and Subbotin 1974) be satisfied, namely:*

$$\min_{v\in V}\max_{u\in U}\left(p_0,\ \phi(T,t,u,v)\right)=\max_{u\in U}\min_{v\in V}\left(p_0,\ \phi(T,t,u,v)\right),0\leq t\leq T.$$

Then in the proof of the theorem 2.1 instead of the set-valued mapping (2.10) one can analyze the set-valued mapping of the form:

$$U_e\left(T,p_0,t\right)=\left\{u,u\in U:\min_{v\in V}\max_{u\in U}\left(p_0,\phi(T,t,u,v)\right)=\min_{v\in V}\left(p_0,\phi(T,t,u,v)\right)\right\}. \tag{2.14}$$

In such case the control $u_t^e(\cdot)$ is independent on $v(t)$.

2.3 INTEGRO-DIFFERENTIAL APPROACH

Let us study the problem of approach for the conflict-controlled process, whose evolution is described by the system of integro-differential equations:

$$\begin{aligned} \dot{z}(t) = A(t)z(t) + \int_{t_0}^{t} K(t,s)z(s)ds + B(t)\,\phi_1(u(t),v(t)) \\ + \int_{t_0}^{t} C(t,s)\phi_2(u(s),v(s))ds + f(t). \end{aligned} \quad (2.15)$$

Here $z(t_0) = z_0, t \geq t_0 \geq 0, z \in R^n$; $A(t), B(t)$ – matrix functions of dimensions $n \times n_1$ and $n \times n_2$, respectively, which are continuous on semi-axis $R_{t_0} = \{t : t \geq t_0\}$; $f(t)$ – continuous on R_{t_0} vector-function; $K(t,s)$, $C(t,s)$ – matrix functions of dimensions $n \times n$, $n \times n_2$, continuous in the closed triangle $\Delta = \{(t,s) : t_0 \leq s \leq t < +\infty\}$; $\phi_1(u,v) : U \times V \to R^{n_1}$, $\phi_2(u,v) : U \times V \to R^{n_2}$ – vector-functions jointly continuous in their variables; U and V – compacts, $U \subset R^{n_1}, V \subset R^{n_2}$.

Controls of the players $u(t)$ and $v(t)$ are measurable functions, taking their values in the sets U and V respectively, that is, $u(\cdot) \in \Omega_U, v(\cdot) \in \Omega_V$. The terminal set of cylindrical form (2.2) is given:

$$M^* = M_0 + M,$$

where M_0 is a linear subspace of R^n, $M \subset coK(L), L = M_0^{\perp}$ – an orthogonal complement to M_0 in R^n.

Our task consists in implementing the technique of subsection 2.4 to the systems of the form (2.15). Note that expression (2.15), in addition, incorporates the integral control unit.

Lemma 2.2. *Let function g(t) be continuous on the half-interval $[t_0, T)$. Then the equation:*

$$\dot{z}(t) = A(t)z(t) + \int_{t_0}^{t} K(t,s)z(s)ds + g(t). \quad (2.16)$$

has a unique solution which is continuous on $[t_0, T)$ and can be written in the form:

$$z(t) = \tilde{g}(t) + \int_{t_0}^{t} \tilde{R}(t,s)\tilde{g}(s)ds,$$

where:

$$\tilde{g}(t) = H(t,t_0)\,z_0 + \int_{t_0}^{t} H(t,\tau)g(\tau)d\tau, \quad (2.17)$$

$H(t,s)$ – *fundamental matrix of homogeneous system:*

$$\dot{z}(t) = A(t)z(t), \quad z(t_0) = z_0, t \in [t_0, T), \quad (2.18)$$

$\tilde{R}(t,s)$ – *the resolvent of matrix $\tilde{K}(t,s)$, appearing as matrix function, continuous on the set $\Delta_T = \{(t,s) : t_0 \leq s \leq t \leq T\}$ and determined by series, uniformly convergent on Δ_T:*

$$\tilde{R}(t,s) = \sum_{n=1}^{\infty} \tilde{K}_n(t,s), \quad (2.19)$$

$$\tilde{K}_1(t,s)=\tilde{K}(t,s),\quad \tilde{K}_n(t,s)=\int_s^t \tilde{K}(t,\tau)\tilde{K}_{n-1}(\tau,s)d\tau,$$
$$n=2,3,\ldots,\quad \tilde{K}(t,s)=\int_s^t H(t,\tau)K(\tau,s)d\tau. \tag{2.20}$$

Proof. A unique continuous solution of equation (2.16) exists by the assumptions on the system (2.15) parameters (Mikhlin 1960). Let us introduce a function:

$$F(t)=\int_{t_0}^t K(t,s)z(s)ds+g(t).$$

It follows from the assumptions of lemma 2.2 that function $F(t)$ is continuous on the half-interval $[t_0,T)$. Let us consider the Cauchy problem for the system:

$$\dot{z}(t)=A(t)z(t)+F(t),\quad z(t)=z_0,\quad t\in[t_0,T).$$

It has a unique solution, which can be presented by the Cauchy formula (Gaishun 2004):

$$z(t)=H(t,t_0)z_0+\int_{t_0}^t H(t,\tau)F(\tau)d\tau.$$

Here $H(t,s)$ is fundamental matrix of the homogeneous system (2.18). Substituting the formula for $F(t)$ into the above expression we obtain:

$$z(t)=H(t,t_0)z_0+\int_{t_0}^t\int_{t_0}^\tau H(t,\tau)K(\tau,s)z(s)\,ds\,d\tau+\int_{t_0}^t H(t,\tau)g(\tau)d\tau.$$

Here from, using the Dirichlet formula (Smirmov 1964), we deduce:

$$z(t)=H(t,t_0)z_0+\int_{t_0}^t\left(\int_{t_0}^\tau H(t,\tau)K(\tau,s)z(s)\,d\tau\right)z(s)ds+\int_{t_0}^t H(t,\tau)g(\tau)d\tau.$$

By (2.17), (2.20) this formula can be rewritten in the form of Volterra equation of second order:

$$z(t)=\tilde{g}(t)+\int_{t_0}^t \tilde{K}(t,s)z(s)ds,\quad t\in[t_0,T), \tag{2.21}$$

where $\tilde{K}(t,s)$, $\tilde{g}(t)$ are functions, continuous in the domains of their definition. Equation (2.21) has a unique solution (Smirmov 1964):

$$z(t)=\tilde{g}(t)+\int_{t_0}^t \tilde{R}(t,s)\tilde{g}(s)ds, t\in[t_0,T),$$

where $\tilde{R}(t,s)$ is given by formulas (2.19), (2.20). The lemma is proved.

Corollary 2.1. *The equation* (2.15) *has a unique (continuous) solution on an arbitrary interval* $[t_0,T]$ *which can be presented in the form:*

$$z(t) = g_0(t) + \int_{t_0}^{t} M(t,s)\phi_1(u(s),v(s))ds + \int_{t_0}^{t} N(t,s)\phi_2(u(s),v(s))ds, \tag{2.22}$$

where:

$$g_0(t) = f_0(t) + \int_{t_0}^{t} \tilde{R}(t,s) f_0(s)\ ds, \tag{2.23}$$

$$f_0(t) = H(t,t_0)z_0 + \int_{t_0}^{t} H(t,s) f(s)\ ds, \quad t \in [t_0, T), \tag{2.24}$$

$$M(t,s) = \tilde{B}(t,s) + \int_{s}^{t} \tilde{R}(t,\tau)\tilde{B}(\tau,s)d\tau, \tag{2.25}$$

$$N(t,s) = \tilde{C}(t,s) + \int_{s}^{t} \tilde{R}(t,\tau)\tilde{C}(\tau,s)d\tau, \tag{2.26}$$

$$\tilde{B}(t,s) = H(t,s)B(s), \tag{2.27}$$

$$\tilde{C}(t,s) = \int_{s}^{t} H(t,\tau)C(\tau,s)d\tau. \tag{2.28}$$

Proof. Lemma 2.2 provides for the solution existence and uniqueness. We set:

$$g(t) = B(t)\phi_1(u(s),v(s)) + \int_{t_0}^{t} C(t,s)\phi_2(u(s),v(s))ds + f(t).$$

Thus, following representation is true:

$$z(t) = \tilde{f}(t) + \int_{t_0}^{t} \tilde{R}(t,s)\tilde{f}(s)ds, \quad t \in [t_0, T).$$

Herewith,

$$\tilde{f}(t) = H(t,t_0)z_0 + \int_{t_0}^{t} H(t,s) f(s)ds + \int_{t_0}^{t} H(t,s)B(s)\phi_1(u(s),v(s))ds$$
$$+ \int_{t_0}^{t}\left(\int_{s}^{t} H(t,\tau)C(\tau,s)d\tau\right)\phi_2(u(s),v(s))ds.$$

Now we rewrite the last formula with account of notation (2.24), (2.27), (2.28):

$$\tilde{f}(t) = f_0(t) + \int_{t_0}^{t} \tilde{B}(t,s)\phi_1(u(s),v(s))ds + \int_{t_0}^{t} \tilde{C}(t,s)\phi_2(u(s),v(s))ds.$$

Then:

$$\int_{t_0}^{t} \tilde{R}(t,s)\tilde{f}(s)ds = \int_{t_0}^{t} \tilde{R}(t,s) f_0(s)ds + \int_{t_0}^{t}\int_{t_0}^{s} \tilde{R}(t,s)\tilde{B}(s,\tau)\phi_1(u(\tau),v(\tau))d\tau\, ds$$
$$+ \int_{t_0}^{t}\int_{t_0}^{s} \tilde{R}(t,s)\tilde{C}(s,\tau)\phi_2(u(\tau),v(\tau))d\tau ds.$$

Let us apply the Dirichlet formula to the two last terms of the above sum and then replace designations τ, s by s, τ, respectively.

$$\int_{t_0}^{t}\int_{t_0}^{s} \tilde{R}(t,s)\,\tilde{B}(s,\tau)\phi_1(u(\tau),v(\tau))\,d\tau\,ds = \int_{t_0}^{t}\left(\int_{s}^{t} \tilde{R}(t,\tau)\,\tilde{B}(\tau,s)\,d\tau\right)\phi_1(u(s),v(s))\,ds,$$

$$\int_{t_0}^{t}\int_{t_0}^{s} \tilde{R}(t,s)\,\tilde{C}(s,\tau)\phi_2(u(\tau),v(\tau))\,d\tau\,ds = \int_{t_0}^{t}\left(\int_{s}^{t} \tilde{R}(t,\tau)\,\tilde{C}(\tau,s)\,d\tau\right)\phi_2(u(s),v(s))\,ds.$$

Considering (2.25), (2.26) we infer formula (2.22). Hence, solution of (2.15) can be presented in the form (2.1). Herewith, $g_0(t)$ is given by (2.23) and is function continuous in $t, t \in R_{t_0}$,

$$\phi(t,s,u,v) = M(t,s)\phi_1(u,v) + N(t,s)\phi_2(u,v),\ (t,s) \in \Delta(t_0),\ u \in U,\ v \in V. \tag{2.29}$$

Function $\phi(t,s,u,v)$ meets all requirements, given in Section 2.3, because $M(t,s), N(t,s)$ are continuous functions in the closed infinite triangle Δ and, by assumption, functions $\phi_1(u,v)$, $\phi_2(u,v)$ are jointly continuous in their variables.

That is why, theorem 2.1 is valid for the game whose dynamics described by the equation (2.15). It should be noted that, in the analyzed game, by the game position is meant the pair $(t, z(T_t, t, u_t(\cdot), v_t(\cdot)))$,

$$z(T_t,t,u_t(\cdot),v_t(\cdot)) = g_0(T) + \int_{t_0}^{T} \phi(T,s,u(s),v(s))\,ds,\ t_0 \le t \le T.$$

Herewith, function $\phi(T,s,u,v)$ is given by formulas (2.29), (2.25), (2.26) and the initial position is the pair:

$$\left(t_0, z(T_0,t_0,u_{t_0}(\cdot),v_{t_0}(\cdot))\right) = \left(t_0, g_0(T_0)\right) = \left(t_0, f_0(T_0) + \int_{t_0}^{T_0} \tilde{R}(T_0,s) f_0(s)\,ds\right).$$

In the above expression $f_0(t)$ is defined by formula (2.24) and $\tilde{R}(t,s)$ – by expressions (2.19), (2.20). In the case under study functions $\lambda(T,t,\zeta)$, $W(T,t,\zeta,p)$, the time of "first absorption" T_0 and the set-valued mapping $W(T, t)$ take the following forms:

$$\lambda(T,t,\zeta) = \min_{\|p\|=1,\ p \in L}\left[W(T,t,\zeta,p) + C(M;-p)\right],$$

$$W(T,t,\zeta,p) = (\zeta,p) + \int_{t}^{T} \min_{v \in V}\max_{u \in U}(\phi(T,s,u,v),p)\,ds,$$

$$t_0 \le t \le T,\quad \zeta \in R^n,\quad p \in R^n.$$

$$T_0 = \min\{T : T \ge t_0, \pi g(T) \in W(T,t_0)\},$$

$$W(T,t_0) = \bigcap_{v(\cdot) \in \Omega_V}\left(M - \int_{t_0}^{t} \pi\phi(T,s,U,v(s))\,ds\right).$$

Thus, with the representation (2.22) for solution of the system of integro-differential equations (2.15), one can easy formulate an analog of the theorem 2.1.

2.4 EXAMPLE OF THE INTEGRO-DIFFERENTIAL GAME APPROACH

Let us consider the game approach problem for the following system:

$$\dot{z}(t)=\lambda z(t)+u(t)-v(t)+\int_0^t (u(s)-v(s))\,ds,\ t_0=0,\ z(0)=z_0,\ z\in R^n. \qquad (2.30)$$

The terminal set M_0 is a linear subspace. In the previous notations $r_1=r_2=n_1=n_2=n$,

$$U=\{u:u\in R^n,\|u\|\le a,a>1\},V=\{v:v\in R^n,\|v\|\le 1\},A(t)=\lambda E,B(t)=E,C(t,s)=E,$$

where E is a unit square matrix of order n, $\phi_1(u,v)=\phi_2(u,v)=u-v$, $f(t)$ is a zero, n-dimensional column vector.

In view of formula (2.15) $K(t,s)$ is a zero square matrix of the order n, therefore, according to (2.20) the same is the matrix $\tilde{K}(t,s)$ with $H(t,s)=e^{\lambda(t-s)}E$. We will analyze two cases.

Case A) $\lambda\ne 0$.

Then, from formulas (2.23), (2.24) there follows that $f_0(t)=g_0(t)=e^{\lambda t}z_0$ and from (2.27), (2.28):

$$\tilde{B}(t,s)=e^{\lambda(t-s)}E,\ \tilde{C}(t,s)=\left(\int_s^t e^{\lambda(t-\tau)}d\tau\right)E=\frac{1}{\lambda}(e^{\lambda(t-s)}-1)E.$$

Using expressions (2.25), (2.26), we find:

$$M(t,s)=e^{\lambda(t-s)}E,\ N(t,s)=\frac{1}{\lambda}(e^{\lambda(t-s)}-1)E.$$

From (2.22) we have got:

$$z(t)=e^{\lambda t}z_0+\int_s^t\left(e^{\lambda(T-s)}\left(1+\frac{1}{\lambda}\right)-\frac{1}{\lambda}\right)(u(s)-v(s))ds,\ t\ge 0.$$

In view of formula (2.29) and expression for $z(T,t,u_t(\cdot),v_t(\cdot))$ one can deduce:

$$\phi(t,s,u,v)=\left(e^{\lambda(t-s)}\left(1+\frac{1}{\lambda}\right)-\frac{1}{\lambda}\right)(u-v),$$

$$\zeta=z(T,t,u_t(\cdot),v_t(\cdot))=e^{\lambda T}z_0+\int_0^t\left(e^{\lambda(t-s)}\left(1+\frac{1}{\lambda}\right)-\frac{1}{\lambda}\right)(u(s)-v(s))ds,\ t\ge 0.$$

Then the function $W(T,t,\zeta,p)$ takes the form:

$$W(T,t,\zeta,p)=(\zeta,p)+\int_t^T\left(\min_{v\in S}\max_{u\in aS}\left(p,\left(e^{\lambda(T-s)}\left(1+\frac{1}{\lambda}\right)-\frac{1}{\lambda}\right)(u-v)\right)\right)ds,\ S=\{v:\|v\|\le 1\}.$$

as:

$$\left(e^{\lambda(T-s)}\left(1+\frac{1}{\lambda}\right)-\frac{1}{\lambda}\right)\ge 0,\ s\in[0,T],$$

for any λ, $\lambda \neq 0$ (it suffice to examine separately cases $\lambda > 0$ and $\lambda < 0$), then:

$$W(T,t,\zeta,p) = (\zeta,p) + (a-1)\|p\|\int_t^T \left(e^{\lambda(T-s)}\left(1+\frac{1}{\lambda}\right) - \frac{1}{\lambda} \right) ds$$
$$= (\zeta,p) + (a-1)\|p\|\frac{1}{\lambda}\left(\left(1+\frac{1}{\lambda}\right)\left(e^{\lambda(T-t)} - 1\right) - (T-t) \right).$$

According to the scheme we obtain:

$$\lambda\left(T,t,z\left(T,t,u_t(\cdot),v_t(\cdot)\right)\right) = \min_{p\in L,\|p\|=1}\left[\left(z\left(T,t,u_t(\cdot),v_t(\cdot)\right), p \right) \right.$$
$$\left. + (a-1)\frac{1}{\lambda}\left(\left(1+\frac{1}{\lambda}\right)\left(e^{\lambda(T-t)} - 1\right) - (T-t) \right) \right]$$
$$= -\left\|\pi z\left(T,t,u_t(\cdot),v_t(\cdot)\right)\right\| + (a-1)\frac{1}{\lambda}\left(\left(1+\frac{1}{\lambda}\right)\left(e^{\lambda(T-t)} - 1\right) - (T-t) \right).$$

By definition, $T_t\left(t,u_t(\cdot),v_t(\cdot)\right)$ is the least root of the following equation in T:

$$\left\|\pi z\left(T,t,u_t(\cdot),v_t(\cdot)\right)\right\| = (a-1)\frac{1}{\lambda}\left(\left(1+\frac{1}{\lambda}\right)\left(e^{\lambda(T-t)} - 1\right) - (T-t) \right), T > t. \tag{2.31}$$

Setting $t = 0$ in (2.31) we obtain the equation for the time of the "first absorption":

$$\left\|\pi e^{\lambda T} z_0\right\| = (a-1)\frac{1}{\lambda}\left(\left(1+\frac{1}{\lambda}\right)\left(e^{\lambda T} - 1\right) - T \right). \tag{2.32}$$

Note that $\|\pi\, z_0\| \neq 0$. Let us rewrite (2.32) in the form:

$$e^{\lambda T}\left((a-1)\frac{1}{\lambda}\left(1+\frac{1}{\lambda}\right) - \|\pi\, z_0\| \right) = (a-1)\frac{1}{\lambda}T + (a-1)\frac{1}{\lambda}\left(1+\frac{1}{\lambda}\right). \tag{2.33}$$

Now we find solution to the following inequality in λ:

$$(a-1)\frac{1}{\lambda}\left(1+\frac{1}{\lambda}\right) - \|\pi\, z_0\| \geq 0.$$

This inequality is equivalent to the inequality:

$$\|\pi z_0\|\lambda^2 - (a-1)\lambda - (a-1) \leq 0,\ \lambda \neq 0,$$

whose solution is described by the inclusion:

$$\lambda \in \left[\frac{a-1-d}{2\|\pi z_0\|}, 0 \right) \cup \left(0, \frac{a-1+d}{2\|\pi z_0\|} \right],\ d = \sqrt{(a-1)^2 + 4\|\pi z_0\|(a-1)}.$$

Solution to the inequality (2.33) has the form:

$$e^{\lambda T}\left(b - \|\pi z_0\|\right) = kT + b,$$
$$b = (a-1)\frac{1}{\lambda}\left(1+\frac{1}{\lambda}\right),\ k = (a-1)\frac{1}{\lambda},\ T > 0. \tag{2.34}$$

With the use of notations:

$$f_1(T)=e^{\lambda T}\left(b-\|\pi z_0\|\right),\ f_2(T)=kT+b,$$

expressions (2.34) can be presented in the form:

$$f_1(T)=f_2(T),\quad T>0. \tag{2.35}$$

Below various possible cases are examined:

1. $\lambda\in\left(0,\dfrac{a-1+d}{2\|\pi z_0\|}\right)$. Then in the equation (2.34) $b-\|\pi z_0\|>0,\ k>0,$ consequently functions $f_1(T),\ f_2(T)$ are monotonically increasing, $f_1(T)$ – as exponential and $f_2(T)$ – as a linear function with $f_1(0)<f_2(0),$ since $f_1(0).=b-\|\pi z_0\|,\ f_2(0)=b.$ That is why there exists a unique root of equation (2.35).
2. $\lambda=\dfrac{a-1+d}{2\|\pi z_0\|}$. Then in (2.34) $b-\|\pi z_0\|=0,\ k>0,\ f_1(T)\equiv 0,\ T\geq 0,\ f_2(0)=b$ and $f_2(T), T>0,$ is monotonically increasing function. In this case equation (2.34) does not have roots.
3. $\lambda>\dfrac{a-1+d}{2\|\pi z_0\|}$. Then in (2.34) $b-\|\pi z_0\|<0,\ k>0,\ f_1(0).=b-\|\pi z_0\|,\ f_2(0)=b,\ f_1(T)$ is monotonically decreasing and $f_2(T)$ – monotonically increasing, respectively. It is evident that in this case equation (2.35) lacks roots too.
4. $\lambda\in\left[\dfrac{a-1-d}{2\|\pi z_0\|},0\right)$. In this case in (2.34): $b-\|\pi z_0\|>0,\ k<0,\ f_1(0)=b-\|\pi z_0\|<b,$ $f_1(0)<f_2(0)=b,\ f_1(T)$ is monotonically decreasing and converging to zero, $f_2(T)$ is monotonically increasing and converging to infinity. Hence, equation (2.35) has a root.
5. $\lambda=\dfrac{a-1-d}{2\|\pi z_0\|}$. Then: $b-\|\pi z_0\|=0,\ k<0,\ f_1(T)\equiv 0,$ function $f_2(T)$ is monotonically increasing, $f_2(0)=b>0.$ Equation (2.35) has a unique root.
6. $\lambda\in\left(-\infty,\dfrac{a-1+d}{2\|\pi z_0\|}\right)$. Then: $b-\|\pi z_0\|<0,\ k<0,$ function $f_1(T)$ is monotonically decreasing and converging to zero, $f_2(T)$ is monotonically increasing and converging to infinity, $f_1(0).=b-\|\pi z_0\|<b=f_2(0).$ Equation (2.35) has a unique root.

Analysis of cases 1–6 shows, that equation (2.32) has a solution, that is, the time of "first absorption" is finite, for:

$$\lambda\in(-\infty,0)\cup\left(0,\frac{a-1+d}{2\|\pi z_0\|}\right).$$

In addition, if $\left\|\pi z\left(T,t,u_t(\cdot),v_t(\cdot)\right)\right\|\neq 0,$ then minimum in the expression for $\lambda\left(T,t,z\left(T,t_1,u_t(\cdot),v_t(\cdot)\right)\right)$ is achieved at the unique element:

$$p_0\left(T,t,z\left(T_0,t_1,u_t(\cdot),v_t(\cdot)\right)\right)=-\frac{1}{\left\|\pi z\left(T,t,u_t(\cdot),v_t(\cdot)\right)\right\|}\pi z\left(T,t,u_t(\cdot),v_t(\cdot)\right).$$

If, otherwise, $\left\|\pi z\left(T,t,u_t(\cdot),v_t(\cdot)\right)\right\|=0,$ minimum is achieved for any $p,\ p\in L.$ Let t be an arbitrary moment of time, at which the game has not been terminated ($z(t)\notin M$). Then, by lemma 2.1 and the definition (2.6), $T_t>t$ and the following inequality has place:

$$(a-1)\frac{1}{\lambda}\left(\left(1+\frac{1}{\lambda}\right)\left(e^{\lambda(T_t-t)}-1\right)-(T_t-t)\right)>0.$$

Since T_t is the root of equation (2.31), $\left\|\pi z\left(T_t,t,u_t(\cdot),v_t(\cdot)\right)\right\|>0$. Because the function $z\left(T,t,u_t(\cdot),v_t(\cdot)\right)$ is t-continuous, there exists a half-interval $[t,t+\delta),\delta>0$, such that $\left\|\pi z\left(T_t,\tau,u_\tau(\cdot),v_\tau(\cdot)\right)\right\|>0,\ \tau\in[t,t+\delta)$. Therefore at $\tau\in[t,t+\delta)$ the set $\Gamma\left(T_t,\tau,z\left(T_t,\tau,u_\tau(\cdot)v_\tau(\cdot)\right)\right)$ consists of a unique element:

$$p_0\left(T_t,\tau,z\left(T_t,\tau,u_t(\cdot),v_\tau(\cdot)\right)\right)=-\frac{1}{\left\|\pi z\left(T_t,\tau,u_\tau(\cdot),v_\tau(\cdot)\right)\right\|}\pi z\left(T_t,\tau,u_\tau(\cdot),v_\tau(\cdot)\right).$$

Thus, it is shown that condition 2) of the theorem 2.1 is fulfilled. On the half-interval $[t,t+\delta)$ the set-valued mapping of the form (2.9) consists of a unique element:

$$u^e\left(T_t,p_0,\tau\right)=ap_0\left(T_t,\tau,z\left(T_t,\tau,u_\tau(\cdot),v_\tau(\cdot)\right)\right),\ \tau\in[t,t+\delta).$$

Case B) $\lambda=0$.

It is evident that then the fundamental matrix is a unit matrix, that is, $H(t,s)=E$. From formulas (2.23), (2.24) we have, that $f_0(t)=z_0, g_0(t)=z_0$, from formulas (2.27), (2.28) we infer $\tilde{C}(t,s)=(t-s)E, B(t,s)=E$ as $B(t)=E, C(t,s)=E$, and from (2.24), (2.25) we have $M(t,s)=E, N(t,s)=(t-s)E$. Then a trajectory of the system trajectory (2.22) in the case (2.30) has a form:

$$z(t)=z_0+\int_0^t(t-s+1)\left(u(s)-v(s)\right)ds,\ t\ge 0.$$

By formula (2.29), $\phi(t,s,u,v)=(t-s+1)(u-v)$ and:

$$z\left(T,t,u_t(\cdot),v_t(\cdot)\right)=z_0+\int_0^t(T-s+1)\left(u(s)-v(s)\right)ds.$$

Then the following formula for function $W(t,s,\zeta;p)$ is true:

$$W(T,t,\zeta,p)=(\zeta,p)+\int_t^T\left(\min_{v\in S}\max_{u\in aS}\left(p,(T-s+1)\right)(u-v)\right)ds.$$

As $T-s+1>0, 0\le t\le s\le T$,

$$\begin{aligned}W(T,t,\zeta,p)&=(\zeta,p)+(a-1)\|p\|\int_t^T(T-s+1)\,ds\\&=(\zeta,p)+(a-1)\|p\|(T-t)\left(\frac{T-t}{2}+1\right).\end{aligned}$$

The from formula (2.31), taking into account that $M=\{0\}$, we obtain:

$$\begin{aligned}\lambda\left(T,t,z\left(T,t,u_t(\cdot),v_t(\cdot)\right)\right)&=\min_{\|p\|=1,p\in L}\left[\left(z\left(T,t,u_t(\cdot),v_t(\cdot)\right),p\right)+(a-1)(T-t)\left(\frac{T-t}{2}+1\right)\right]\\&=-\left\|\pi z\left(T,t,u_t(\cdot),v_t(\cdot)\right)\right\|+(a-1)(T-t)\left(\frac{T-t}{2}+1\right).\end{aligned}$$

According to (2.31) $T\left(t,u_t(\cdot),v_t(\cdot)\right)$ is defined as the least root of the following equation in T:

$$(a-1)(T-t)\left(\frac{T-t}{2}+1\right)=\left\|\pi z\left(T,t,u_t(\cdot),v_t(\cdot)\right)\right\|,\ T\ge t.$$

Hence, the time of "first absorption" T_0 appears as the least positive root of the equation:

$$(a-1)T\left(\frac{T}{2}+1\right)=\|\pi z_0\|,\ T>0.$$

It is easy to see that:

$$T_0=-1+\sqrt{1+\frac{2\|\pi z_0\|}{a-1}}.$$

Evidently, the time of "first absorption" is finite, regardless of the initial state of the game $(z_0, \|\pi z_0\| < \infty)$. Analogously to the case *A*) $\lambda \neq 0$, it can be shown that the condition 2) of theorem 2.1 is fulfilled and therefore the sought positional control is determined. Combining the cases *A*) and *B*), we come to the following conclusion. If:

$$\lambda \in \left(-\infty, \frac{a-1+d}{2\|\pi z_0\|}\right), d=\sqrt{(a-1)^2+4\|\pi z_0\|(a-1)},$$

then approaching the terminal set can be performed in the class of positional controls at the time, less or equal than the "first absorption" time.

2.5 QUASILINEAR POSITIONAL INTEGRAL GAMES APPROACH

Let us consider the conflict-controlled process, governed by the system of linear integral Volterra equations of the second kind:

$$z(t)=f(t)+\int_0^t \bar{K}(t,s)z(s)ds+\int_0^t Q(t,s)\varphi(u(s),v(s))ds, t\geq 0. \tag{2.36}$$

Here $z \in R^n$; $f(t)$ is a vector-function, continuous on the positive semi-axis $R_+ = \{t: t\geq 0\}$; $\bar{K}(t,s), Q(t,s)$ are $n\times n$ and $n\times r$ matrices continuous on the triangle:

$$\Delta=\{(t,s): 0\leq s\leq t<+\infty\}.$$

Control domains are compacts U, V, $U\subset R^{n_1}, V\subset R^{n_2}$, vector-function $\phi(u,v)$, $\phi: U\times V \to R^r$, is jointly continuous in its variables.

The terminal set has a cylindrical form:

$$M^* = M_0 + M,$$

where M_0 is a linear subspace in R^n, $M \in coK(L)\ L = M_0^{\perp}$.

Admissible controls of the players are supposed to be measurable functions. The goals of the players are the same as in the previous section. We assume that $z(0)=f(0)\notin M^*$.

Lemma 2.2. *Let elements of the matrix* $\bar{K}(t,s)$ *be continuous on the closed triangle* Δ *and function* $g(t)$ *– continuous on the closed interval* [0, *T*]. *Then the equation:*

$$z(t)=g(t)+\int_0^t \bar{K}(t,s)z(s)ds, t\in[0,T],$$

has a unique solution of the form:

$$z(t) = g(t) + \int_0^t R(t,s)g(s)ds, \; t \in [0,T].$$

Here $R(t,s)$ is the matrix $\bar{K}(t,s)$ resolvent, which is a matrix function, continuous on Δ_T and defined by the following, converging on Δ series:

$$R(t,s) = \sum_{n=1}^{\infty} \bar{K}_n(\tau,s), \; \bar{K}_1(t,s) = \bar{K}(t,s), \; \bar{K}_n(t,s) = \int_s^t \bar{K}(t,\tau)\bar{K}_{n-1}(\tau,s)d\tau, \; n = 2,3,\ldots \quad (2.37)$$

Corollary 2.2. *Equation* (2.36) *has a unique continuous solution on arbitrary interval* $[0,T]$, $T > 0$:

$$z(t) = f(t) + \int_0^t R(t,s)f(s)ds + \int_0^t N_1(t,s)\phi(u(s),v(s))ds \quad (2.38)$$

Here $N_1(t,s)$,

$$N_1(t,s) = Q(t,s) + \int_s^t R(t,\tau)Q(\tau,s)ds, \quad (2.39)$$

is a continuous matrix function.

Proof. The first statement immediately follows from lemma 2.2. Then, by the same lemma, since:

$$g(t) = f(t) + \int_0^t Q(t,s)\phi(u(s),v(s))ds,$$

we have:

$$\begin{aligned} z(t) &= f(t) + \int_0^t Q(t,s)\phi(u(s),v(s))ds \\ &+ \int_0^t R(t,s)\left(f(s) + \int_0^s Q(s,\tau)\phi(u(\tau),v(\tau))d\tau\right)ds \\ &= f(t) + \int_0^t R(t,s)f(s)ds + \int_0^t Q(t,s)\phi(u(s),v(s))ds \\ &+ \int_0^t \int_0^s R(t,s)\, Q(s,\tau)\phi(u(\tau),v(\tau))d\tau ds. \end{aligned}$$

Let us apply the Dirichlet formula in the last term. Then, successively replacing notations τ for s and s for τ, we obtain:

$$\begin{aligned} z(t) &= f(t) + \int_0^t R(t,s)f(s)ds + \int_0^t Q(t,s)\phi(u(s),v(s))ds \\ &+ \int_0^t \int_\tau^t R(t,s)Q(s,\tau)\phi(u(\tau),v(\tau))ds\,d\tau \\ &= f(t) + \int_0^t R(t,s)f(s)ds + \int_0^t Q(t,s)\phi(u(s),v(s))ds \\ &+ \int_0^t \left(\int_s^t R(t,\tau)\, Q(\tau,s)\, d\tau\right)\phi(u(s),v(s))ds \end{aligned}$$

$$= f(t) + \int_0^t R(t,s) f(s) ds + \int_0^t \left(Q(t,s) + \int_s^t R(t,\tau) Q(\tau,s)\, d\tau \right) \phi(u(s),v(s)) ds$$
$$= f(t) + \int_0^t R(t,s)\, f(s) ds + \int_0^t N_1(t,s) \phi(u(s),v(s)) ds.$$

Thus, solution to the system of integral equations (2.36) can be presented in the form (2.1), where $t_0 = 0$,

$$g(t) = f(t) + \int_0^t R(t,s)\, f(s) ds,\ \phi(t,s,u,v) = N_1(t,s) \phi(u,v). \tag{2.40}$$

Now we provide examples of resolvents for some types of continuous on Δ_1 kernels obtained from representation (2.37). Let Γ be a square matrix of order n and e^A be a matrix exponent. Then following assertions are true:

1. if $\bar{K}(t,s) = A$, then $R(t,s) = Ae^{(t-s)A}$;
2. if $\bar{K}(t,s) = (t-s)A$, then $R(t,s) = \sum_{n=1}^{\infty} \frac{(t-s)^{2n-1}}{(2n-1)!} A^n$,

2.1 e. g. in the case $A = E$, $R(t,s) = sh(t-s)E$, $sht = \frac{e^t - e^{-t}}{2} = \sum_{n=1}^{\infty} \frac{t^{2n-1}}{(2n-1)!}$;

3. if $\bar{K}(t,s) = e^{(t-s)A}$, then $R(t,s) = e^{t-s} Ae^{(t-s)A}$;
4. if $\bar{K}(t,s) = \alpha e^{\beta(t-s)} e^{\gamma(t-s)A}$, where α, β, γ are real numbers, then $R(t,s) = \alpha\, e^{(\alpha+\beta)(t-s)+\gamma(t-s)A}$.

Since solution to (2.38) can be presented in the form (2.1) we can apply construction, outlined in the section 2.3. Evidently theorem 2.1 is also valid in this case. Considering the notations of Section 2.4 we obtain:

$$z(T,t,\, u_t(\cdot),\, v_t(\cdot)) = z_T^0 + \int_0^t N(T,s) \phi(u(s),v(s)) ds. \tag{2.41}$$

Here:

$$z_T^0 = f(T) + \int_0^T R(T,s) f(s) ds. \tag{2.42}$$

It is easy to see, that for the process evolving according to (2.36),

$$\begin{aligned}
&\zeta = z(T,t,\, u_t(\cdot),\, v_t(\cdot)),\\
&W(T,t,\zeta;p) = (\zeta,p) + \int_t^T \min_{v\in V} \max_{u\in U} (p, N_1(t,s)\phi(u,v)) ds,\\
&\lambda(T,t,\zeta) = \min_{\|p\|=1,\, p\in L} \left[W(T,t,\zeta;p) + C(M;-p) \right],\\
&W(t) = \bigcap_{v(\cdot)\in\Omega_V} \left(M - \int_0^t \pi\, N_1(t,s) \phi(U,v(s)) ds \right),\\
&T_0 = \min\{t \geq 0 : \pi\, g(t) \in W(t)\} = \min\{t \geq 0 : \lambda(t,0,g(t))\}.
\end{aligned} \tag{2.43}$$

We see that the integral game (2.36) is a specific case of the game (2.1). Then the analog of the theorem 2.1 is true, under the conditions for the non-emptity of $W(t)$ $(W(t) \neq \varnothing, \forall t \geq 0)$ and the uniqueness of extremal element in the expression defining $\lambda(T,t,\zeta)$.

Remark 2.2. If $\phi(u,v) = -u + v$, then:

$$W(t) = \left(M + \int_0^t \pi N_1(t,s)Uds \right) \underline{*} \int_0^t \pi N_1(t,s)Vds.$$

2.6 CASE OF THE PURSUIT PROBLEM

Here we examine the dynamic system with separated motions of the players. Let motions of the players at $t \in [0,+\infty)$ be described, respectively, by the second-kind linear integral Volterra equations:

$$x(t) = f_1(t) + \int_0^t A_1(t,s)x(s)ds + \int_0^t B_1(t,s)u(s)ds, \tag{2.44}$$

$$y(t) = f_2(t) + \int_0^t A_2(t,s)y(s)ds + \int_0^t B_2(t,s)v(s)ds. \tag{2.45}$$

Here $x \in R^{n_1}, y \in R^{n_2}, f_1(t)$ and $f_2(t), t \in [0,+\infty)$, – n_1 - and n_2 - dimensional continuous functions, $A_1(t,s), B_1(t,s), A_2(t,s), B_2(t,s)$ – matrices of orders $n_1 \times n_1, n_1 \times r_1, n_2 \times n_2$, and $n_2 \times r_2$, respectively, which are continuous on the closed infinite triangle $\Delta = \{(t,s) : 0 \le s \le t < +\infty\}$. Controls of the players – $u(t)$ and $v(t)$ are, respectively, r_1 - and r_2 - measurable vector-functions, chosen in Ω_U, Ω_V.

By virtue of corollary of lemma 2.2 and assumptions concerning the equations (2.44) and (2.45) right-hand sides there exist unique continuous solutions $(x(t), y(t))$ to these equations on an arbitrary interval $[0,t], t > 0$. It is assumed that at the initial moment of time $t = 0$:

$$\left\| \{x(0)\}_m - \{y(0)\}_m \right\| = \left\| \{f_1(0)\}_m - \{f_2(0)\}_m \right\| > \varepsilon,$$

where $\{x\}_m$ is a vector consisting of the first m coordinates of vector x, $m < \min(n_1, n_2)$, and ε – given number, $\varepsilon > 0$.

The task of the first player is to choose own admissible control in a way the following inequality be satisfied in the shortest time:

$$\left\| \{x(T)\}_m - \{y(T)\}_m \right\| \le \varepsilon,$$

under any arbitrary control of the second player. Now we introduce the notations:

$$z(t) = \begin{pmatrix} x(t) \\ y(t) \end{pmatrix}, \quad f(t) = \begin{pmatrix} f_1(t) \\ f_2(t) \end{pmatrix}, \quad A(t,s) = \begin{pmatrix} A_1(t,s) & O_1 \\ O_2 & A_2(t,s) \end{pmatrix},$$

$$Q(t,s) = \begin{pmatrix} B_1(t,s) & O_1 \\ O_2 & B_2(t,s) \end{pmatrix}, \quad \phi(u,v) : U \times V \to R^{n_1+n_2}.$$

Here O_1, O_2, O_3, O_4 – zero matrices of the dimensions $n_1 \times n_2$, $n_2 \times n_1$, $n_1 \times r_2$, $n_2 \times r_1$, respectively.

Thus, we come to the approach problem, examined in the previous subsection, with the terminal set $M^* = M_0 + M$, where:

$$M_0 = \begin{pmatrix} a \\ a_1 \\ a \\ a_2 \end{pmatrix}, \quad a \in R^m, \ a_1 \in R^{n_1-m}, \ a_2 \in R^{n_2-m}, \tag{2.46}$$

$$M=\begin{pmatrix} p \\ o_1 \\ -p \\ o_2 \end{pmatrix},\ p\in R^m,\ \|p\|\le\frac{\varepsilon}{2},\ o_1\in R^{n_1-m},\ o_2\in R^{n_2-m}. \tag{2.47}$$

Here o_1, o_2 are zero vectors.

Therefore, to solve the problem under study Theorem 2.1 can be applied. Let us denote for $0<t\le T$:

$$x(T,t,u_t(\cdot))=f_1(T)+\int_0^T R_1(T,s)\,ds+\int_0^t N_1(T,s)u(s)\,ds,$$
$$y(T,t,v_t(\cdot))=f_2(T)+\int_0^T R_2(T,s)\,ds+\int_0^t N_2(T,s)v(s)\,ds.$$

Here:

$$N_1(t,s)=B_1(t,s)+\int_s^t R_1(t,\tau)B_1(\tau,s)\,ds,\ N_2(t,s)=B_2(t,s)+\int_s^t R_2(t,\tau)B_2(\tau,s)\,ds,$$

where $R_1(t,s)$, $R_2(t,s)$ are resolvents of the matrices $A_1(t,s)$, $A_2(t,s)$. In the new notations:

$$z(T,t,u_t(\cdot),v_t(\cdot))=\begin{pmatrix} x(T,t,u_t(\cdot)) \\ y(T,t,v_t(\cdot)) \end{pmatrix}=\begin{pmatrix} f_1(T)+\int_0^T R_1(T,s)\,ds+\int_0^t N_1(T,s)u(s)\,ds \\ f_2(T)+\int_0^T R_2(T,s)\,ds+\int_0^t N_2(T,s)v(s)\,ds \end{pmatrix}. \tag{2.48}$$

The function $\lambda(T,t,z(T,t,u_t(\cdot),v_t(\cdot)))$, with account of notations (2.47), (2.48), takes the form:

$$\lambda(T,t,z(T,t,u_t(\cdot),v_t(\cdot)))=\frac{1}{\sqrt{2}}\min_{p\in R^m,\|p\|=1}\Big(\big(p,\{x(T,t,u_t(\cdot))\}_m-\{y(T,t,v_t(\cdot))\}_m\big)$$
$$+\int_t^T\max_{u\in U}\big(p,\{N_1(T,s)\,u\}_m\big)\,ds-\int_t^T\max_{v\in V}\big(p,\{N_2(T,s)\,v\}_m\big)\,ds+\varepsilon\Big).$$

Let us denote:

$$\lambda(T,t,x(T,t,u_t(\cdot)),y(T,t,v_t(\cdot)))=\min_{p\in R^n,\|p\|=1}\Big(\big(p,\{x(T,t,u_t(\cdot))\}_m-\{y(T,t,v_t(\cdot))\}_m$$
$$+\int_t^T\max_{u\in U}\big(p,\{N_1(T,s)\,u\}_m\big)\,ds-\int_t^T\max_{v\in V}\big(p,\{N_2(T,s)\,v\}_m\big)\,ds+\varepsilon\Big).$$

$$T(t,u_t(\cdot),v_t(\cdot))=\min\{T\ge t:\lambda(T,t,x(T,t,u_t(\cdot)),y(T,t,v_t(\cdot)))=0\},$$
$$T_0=\min\{T\ge 0:\lambda(T,0,x(T,0,\cdot,\cdot),y(T,0,\cdot,\cdot))=0\},$$
$$\Gamma(T,t,x(T,t,u_t(\cdot)),y(T,t,v_t(\cdot)))=\{p:p\in R^m,\|p\|=1,W(T,t,x,y;p)+\varepsilon=\lambda(T,t,x,y)\},$$
$$\tilde{W}(T,t,x,y;p)=\big(p,\{x\}_m-\{y\}_m\big)$$
$$+\int_t^T\max_{u\in U}\big(p,\{N_1(T,s)\,u\}_m\big)\,ds-\int_t^T\max_{v\in V}\big(p,\{N_2(T,s)v\}_m\big)\,ds,$$
$$T_t=T(t,u_t(\cdot),v_t(\cdot)),$$

where:

$$x = x\left(T,t,u_t(\cdot)\right), \quad y = y\left(T,t,v_t(\cdot)\right).$$

In what follows the following triple will be referred to as the game position:

$$\left(t, x\left(T,t,u_t(\cdot)\right), y\left(T,t,v_t(\cdot)\right)\right).$$

Theorem 2.2. *Let in the game of approach for dynamic system with distributed motions of the players* (2.44), (2.45):

1. *the "first absorption" time T_0 is finite;*
2. *for any position $\left(t^*, x\left(T_{t^*},t^*, u_{t^*}(\cdot)\right), y\left(T_{t^*},t^*, v_{t^*}(\cdot)\right)\right) \in R^{n_1+n_2+1}, 0 \le t^* \le T_0, T_{t*} \le T_0$, there exists a neighborhood $\Omega\left(t^*, x\left(T_{t^*},t^*, u_{t^*}(\cdot)\right), y\left(T_{t^*},t, v_{t^*}(\cdot)\right)\right)$, such that for all $(t,\chi,\gamma) \in \Omega$ $\left(t^*, x\left(T_{t^*},t^*, u_{t^*}(\cdot)\right), y\left(T_{t^*},t, v_{t^*}(\cdot)\right)\right)$, $\chi \in R^{n_1}, \gamma \in R^{n_2}$, the set $\Gamma\left(T_{t^*},t, \chi,\gamma\right)$ consists of a unique element $p_0\left(T_{t^*},t, \chi,\gamma\right)$.*

Then the first player can terminate the game no later than at the time T_0 for any admissible control of the second player.

The proof is analogous to that of the theorem 2.1. Here we only note that control of the first player on the half-interval $[t_k,t_{k+1}), k = 0,1,2,\ldots, \quad t_0 = 0$, is to be constructed in the form of measurable selection $u^e(t)$ of the set-valued mapping:

$$\begin{aligned} U^e\left(T_{t_k}, p_0\left(T_{t_k},t,\chi,\gamma\right),t\right) &= \left\{u : u \in U, \max_{u \in U}\left(p_0\left(T_{t_k},t,\chi,\gamma\right), \left\{N_1\left(T_{t_k},t\right)u\right\}_m\right)\right\} \\ &= \left(p_0\left(T_{t_k},t,\chi,\gamma\right), \left\{N_1\left(T_{t_k},t\right)u\right\}_m\right), t \in [t_k,t_{k+1}), \\ \chi &= x\left(T_{t_k},t, u_t^e(\cdot)\right); \quad \gamma = x\left(T_{t_k},t, v_t(\cdot)\right). \end{aligned}$$

2.7 CONNECTION OF INTEGRAL AND DIFFERENTIAL GAMES OF APPROACH

Let us explore the nonstationary differential game of approach with the dynamics:

$$\dot{z}(t) = A(t)z(t) + \phi\left(u(t),v(t)\right), \; t \ge 0, \; z(t) \in R^n, \; z(0) = z_0. \tag{2.49}$$

Here $\phi(u,v): U \times V \to R^n$ is vector-function, jointly continuous in its variables.

It is evident how this game can be reduced to the integral one. Integrating both sides of the equation (2.49) over the interval $[0,t]$, we obtain:

$$z(t) = z_0 + \int_0^t A(s)z(s)ds + \int_0^t \phi\left(u(s),v(s)\right)ds, \quad t \ge 0. \tag{2.50}$$

One can see that equation (2.50) is a particular case of the equation (2.36):

$$z(t) = f(t) + \int_0^t \bar{K}(t,s)z(s)ds + \int_0^t Q(t,s)\phi\left(u(s),v(s)\right)ds, \tag{2.50}$$

where $f(t) = z_0; \bar{K}(t,s) = A(s), Q(t,s) = E$ (E is a unit matrix of order n).

Then, the technique presented in the Section 2.4, can be applied to analyze the integral approach game (2.50). Let us return to equation (2.36) and set $f(t) = z_0, t \ge 0$,

$\bar{K}(t,s)=A, Q(t,s)=E$, where A is a square $n\times n$ matrix. Then, as one can observe, (2.36) turns into a specific case of the quasilinear differential game, namely:

$$\dot{z}=Az+\phi(u,v),\ z(0)=z_0.$$

In view of formulas (2.37), (2.39), here $R(t,s)=Ae^{(t-s)A}$, $N_1(t,s)=e^{(t-s)A}$. Then, after the separation of the terms, containing the controls $u^0(s), v^0(s), s\in[0,t]$, equation (2.28) takes the form:

$$\begin{aligned}
z(T) &= f(T)+\int_0^T R(T,s)f(s)ds+\int_0^t N_1(T,s)\phi(u^0(s),v^0(s))ds\\
&+\int_t^T N_1(T,s)\phi(u(s),v(s))ds\\
&= z_0+\int_0^T Ae^{(T-s)A}z_0\,ds+\int_0^t e^{(T-s)A}\,\phi(u^0(s),v^0(s))ds+\int_t^T e^{(T-s)A}\,\phi(u(s),v(s))ds\\
&= e^{(T-t)A}\left(e^{tA}z_0+\int_0^t e^{(t-s)A}\phi(u^0(s),v^0(s))ds\right)\\
&+\int_t^T e^{(T-s)A}\phi(u(s),v(s))ds = e^{(T-t)A}z(t)+\int_t^T e^{(T-s)A}\phi(u(s),v(s))ds\\
&= e^{(T-t)A}z(t)+\int_t^T e^{(T-s)A}\phi(u(s),v(s))ds\\
&= e^{(T-t)A}z(t)+\int_0^{T-t} e^{(T-t-s)A}\phi(u(s+t),v(s+t))ds.
\end{aligned}$$

Therefore, in the case of quasilinear differential game the position $z(t)$ involves information on the controls, used on the half-interval $[0,t)$. Thus, if the players apply controls $u_t^0(s), v_t^0(s)$ on $[0,t)$, knowledge of the state $z(t)$ is sufficient to define all possible states before $T, T\ge t$. This is an important point to clarify relation of the above constructions with the method of B.N. Pshenichnyi (1968).

Below obtained results are demonstrated on the Pontryagin's Testing Example. Let motions of the pursuer and the evader be given by the equations:

$$\ddot{x}(t)+\alpha\dot{x}(t)=\rho u, \tag{2.51}$$

$$\ddot{y}(t)+\beta\dot{y}(t)=\sigma v, \tag{2.52}$$

$$t\ge 0,\ x(t)\in R^n, n\ge 2,\ y(t)\in R^n, \alpha>0, \beta>0, \rho>0, \sigma>0.$$

The players choose their controls in a way their realizations belong to the sets of admissible controls $\Omega_U, \Omega_V, U=V=S$.

The initial states and velocities of the players are given:

$$x(0)=x_0,\ \dot{x}(0)=\dot{x}_0,\ y(0)=y_0,\ \dot{y}(0)=\dot{y}_0,\ x_0\ne y_0.$$

The game is considered to terminate when geometric coordinates of the players coincide, that is, $x=y$. With the help of substitution:

$$z_1=x,\ z_2=\dot{x},\ z_3=y,\ z_4=\dot{y},$$

equations (2.51) and (2.52) are reduced to the following system of equations:

$$\dot{z}(t) = Az(t) + Bu + Cv, \tag{2.53}$$

with the initial conditions:

$$z(0) = z_0 = (x_0, \dot{x}_0, y_0, \dot{y}_0),$$

where:

$$A = \begin{pmatrix} 0 & 1 & 0 & 0 \\ 0 & -\alpha & 0 & 0 \\ 0 & 0 & 0 & 1 \\ 0 & 0 & 0 & -\beta \end{pmatrix} \otimes E_n; \quad B = \begin{pmatrix} 0 \\ \rho \\ 0 \\ 0 \end{pmatrix} \otimes E_n; \quad C = \begin{pmatrix} 0 \\ 0 \\ 0 \\ \sigma \end{pmatrix} \otimes E_n. \tag{2.54}$$

Here E_n is a unit $n \times n$ matrix, sign $\otimes$ – the Kronecker product (Gantmakher 1987). The pursuer strives in the shortest time to achieve the coincidence of his geometric coordinates with the geometric coordinates of the evader. Thus, the terminal set has the form:

$$M_0 = \{(x, y, \dot{x}, \dot{y}) : x = y\}.$$

One can see that M_0 presents itself a linear subspace in R^{4n}. We assume that $z_0 \notin M_0$. Let us reduce the original game to the equivalent integral game in the way described at the beginning of this subsection. Then we have:

$$z(t) = z_0 + \int_0^t Az(s)ds + \int_0^t Bu(s)ds + \int_0^t Cv(s)ds. \tag{2.55}$$

Now we apply the method, provided in section 2.3. Evidently, the barrier cone of the set M has the form:

$$K_{M_0} = \{ (p, 0, -p, 0), p \in R^n \} = L.$$

By symbol 0 is denoted zero vector belonging to R^n. Using corollary of lemma 2.2 we obtain the solution to the equivalent systems (2.53). (2.55) in the form of Cauhy formula:

$$z(t) = e^{tA} z_0 + \int_0^t e^{(t-s)A} Bu(s)ds + \int_0^t e^{(t-s)A} Cv(s)ds.$$

Let us adapt the technique developed for investigation the positional integral games to the case under study. Here:

$$z(T, t, u_t(\cdot), v_t(\cdot)) = e^{TA} z_0 + \int_0^t e^{(T-s)A} Bu(s)ds + \int_0^t e^{(T-s)A} Cv(s)ds$$
$$= e^{(T-t)A} \left(e^{tA} z_0 + \int_0^t e^{(t-s)A} Bu(s)ds + \int_0^t e^{(t-s)A} Cv(s)ds \right) = e^{(T-t)A} z(t).$$

Since:

$$C(M_0; l) = \begin{cases} 0, & l \in K_{M_0} \\ +\infty, & l \notin K_{M_0} \end{cases},$$

then:

$$\lambda\left(T,t,z\left(T,t,u_t(\cdot)\right),v_t(\cdot)\right)$$
$$= \min_{\|l\|=1,\, l\in K_{M_0}} \left(\left(l, e^{(T-t)A}z(t)\right) + \int_0^T \max_{u\in U}\left(l, e^{sA}Bu\right)ds + \int_0^T \min_{v\in V}\left(l, e^{sA}Cv\right)ds \right).$$

After substitution into this formula explicit expressions for l and e^{tA}:

$$l=(p,0,-p,0),\, l\in L,\, e^{tA}=\begin{pmatrix} 1 & \frac{1}{\alpha}\left(1-e^{-\alpha t}\right) & 0 & 0 \\ 0 & e^{-\alpha t} & 0 & 0 \\ 0 & 0 & 1 & \frac{1}{\beta}\left(1-e^{-\beta t}\right) \\ 0 & 0 & 0 & e^{-\beta t} \end{pmatrix} \otimes E_n,$$

we obtain:

$$\begin{aligned}
&\lambda\left(T,t,z\left(T,t,u_t(\cdot)\right),v_t(\cdot)\right)\\
&= \min_{\|p\|=\frac{1}{\sqrt{2}},\, p\in R^n} \left(\left(p, z_1(t) - z_3(t) + \frac{1}{\alpha}\left(1-e^{-\alpha(T-t)}\right) z_2(t) - \frac{1}{\beta}\left(1-e^{-\beta(T-t)}\right) z_4(t) \right) \right.\\
&\left. + \int_t^T \frac{\rho}{\alpha}\left(1-e^{-\alpha s}\right) \max_{u\in S}(p,u)\, ds + \int_t^T \frac{\sigma}{\beta}\left(1-e^{-\beta s}\right) \max_{v\in S}(-p,v)\, ds \right)\\
&= \min_{\|p\|=\frac{1}{\sqrt{2}},\, p\in R^n} \left(\left(p, z_1(t) - z_3(t) + \frac{1}{\alpha}\left(1-e^{-\alpha(T-t)}\right) z_2(t) - \frac{1}{\beta}\left(1-e^{-\beta(T-t)}\right) z_4(t) \right) \right.\\
&\left. + \left(\int_t^T \frac{\rho}{\alpha}\left(1-e^{-\alpha s}\right) ds + \int_t^T \frac{\sigma}{\beta}\left(1-e^{-\beta s}\right) ds \right) \|p\| \right) \qquad (2.56)\\
&= \frac{1}{\sqrt{2}} \min_{\|p\|=1,\, p\in R^n} \left(\left(p, z_1(t) - z_3(t) + \frac{1}{\alpha}\left(1-e^{-\alpha(T-t)}\right) z_2(t) - \frac{1}{\beta}\left(1-e^{-\beta(T-t)}\right) z_4(t) \right) \right.\\
&\left. + \frac{\rho}{\alpha}\left(T - t + \frac{1}{\alpha}e^{-\alpha T} - \frac{1}{\alpha}e^{-\alpha t} \right) - \frac{\sigma}{\beta}\left(T - t + \frac{1}{\beta}e^{-\beta T} - \frac{1}{\beta}e^{-\beta t} \right) \right)\\
&= \frac{1}{\sqrt{2}} \left(- \left\| z_1(t) - z_3(t) + \frac{1}{\alpha}\left(1-e^{-\alpha(T-t)}\right) z_2(t) - \frac{1}{\beta}\left(1-e^{-\beta(T-t)}\right) z_4(t) \right\| \right.\\
&\left. + \frac{\rho}{\alpha}\left(T - t + \frac{1}{\alpha}e^{-\alpha T} - \frac{1}{\alpha}e^{-\alpha t} \right) - \frac{\sigma}{\beta}\left(T - t + \frac{1}{\beta}e^{-\beta T} - \frac{1}{\beta}e^{-\beta t} \right) \right).
\end{aligned}$$

Now we set $t=0$ and find the time of "first absorption" T_0 as the first positive root of the following equation in T:

$$\begin{aligned}
&\left\| x_0 - y_0 + \frac{1}{\alpha}\left(1-e^{-\alpha(T-t)}\right)\dot{x}_0 - \frac{1}{\beta}\left(1-e^{-\beta(T-t)}\right)\dot{y}_0 \right\|\\
&= \frac{\rho}{\alpha}\left(T - \frac{1}{\alpha}\left(1-e^{-\alpha T}\right) \right) - \frac{\sigma}{\beta}\left(T - \frac{1}{\beta}\left(1-e^{-\beta T}\right) \right).
\end{aligned} \qquad (2.57)$$

Here the set $\Gamma(T_t,t,z)$ consists of a unique element:

$$l_0(T_t,\tau,z) = -\frac{1}{\|\pi z\|}\pi z, \|\pi z\| \neq 0.$$

$T_t = T_t(t,u_t(\cdot),v_t(\cdot))$, $z = z(T_t,\tau,u_\tau(\cdot),v_\tau(\cdot))$, $\tau \in [t,t+\delta)$, $\delta > 0$, t – an arbitrary moment of time, $0 \le t \le T_0$, π – the operator of orthogonal projection from R^{4n} onto K_{M_0}, having the form:

$$\pi = \frac{1}{2}\begin{pmatrix} 1 & 0 & -1 & 0 \\ 0 & 0 & 0 & 0 \\ -1 & 0 & 1 & 0 \\ 0 & 0 & 0 & 0 \end{pmatrix} \otimes E_n.$$

Taking into account that:

$$z = e^{(T_t-\tau)A}z(\tau),$$

we deduce:

$$l_0(T_t,\tau,z) = l_0\left(T_t,\tau,e^{(T_t-\tau)A}z(\tau)\right)$$
$$= -\frac{1}{\sqrt{2}\|p(T_t,\tau,z(\tau))\|}\left(p(T_t,\tau,z(\tau)),o,-p(T_t,\tau,z(\tau)),o\right),$$

where:

$$p(T_t,\tau,z(\tau)) = z_1(\tau) - z_3(\tau) + \frac{1}{\alpha}\left(1-e^{-\alpha(T_t-\tau)}\right)z_2(\tau) - \frac{1}{\beta}\left(1-e^{-\beta(T_t-\tau)}\right)z_4(\tau).$$

Further considerations lead to the results analogous to those presented in (Pshenichnyi 1969). We give them without proof. Namely, the assumptions of the theorem 2.1 are fulfilled if following conditions on the game parameters are fulfilled, one of the inequalities to be strictly satisfied:

$$\rho \ge \sigma, \quad \frac{\rho}{\alpha} \ge \frac{\sigma}{\beta}.$$

Note that at $t = 0$ the left-hand side of the equation (2.57) is equal to $\|x_0 - y_0\|$, while the right-hand side turns into zero. Besides, as $t \to +\infty$ the former has bounded value while the latter is linearly converging to infinity, by virtue of the condition $\frac{\rho}{\alpha} \ge \frac{\sigma}{\beta}$.

Since both sides of equation (2.56) are continuous functions, the time of "first absorption" is finite for any initial position. That is why, the control of the pursuer, realizing the game goal, is uniquely defined, namely,

$$u^e(T_t,\tau,z) = u^e\left(T_t,\tau,e^{(T_t-\tau)A}z(\tau)\right) = -\frac{1}{\|p(T_t,\tau,z(\tau))\|}p(T_t,\tau,z(\tau)).$$

2.8 POSITIONAL CONTROL IN DIFFERENTIAL-DIFFERENCE GAMES

Let the dynamics of a conflict-controlled process be described by the system of difference-differential equations (Bellman and Cooke 1963, Matychyn and Chikrii A.A. and Gromaszek 2012):

$$\dot{z}(t) = Az(t) + Bz(t-\tau) - \phi(u,v), \tag{2.58}$$

where $z \in R^n, u \in U, v \in V$, A is a square matrix of order n, u and v – control parameters, chosen by the pursuer and the evader from the compacts U and V, respectively, $U \subset R^n$, $V \subset R^n$, $\phi(u, v)$ – function, jointly continuous in its variables. The terminal set of cylindrical type is given:

$$M^* = M_0 + M,$$

where M_0 – a linear subspace of R^n, $M \in coK(L)$, $L = M_0^{\perp}$.

The absolutely continuous function $z^0(t)$, defined on the interval $t \in [-\tau, 0]$, is given, which stands for the initial state of the process. The state at current moment of time t, $t \in [0, +\infty)$, is determined by the vector-function $z^t(s)$, $z^t(s) = z(t+s)$, $-\tau \le s \le 0$, which is the portion of the process trajectory on the time-interval $[t-\tau, t]$.

The goal of the pursuer is in the shortest time to bring a trajectory of the system (2.58) to the terminal set M. The game is considered terminated at the moment T, $T > 0$, such that $z(T) \in M$ or $\pi z(T) \in L$. Controls $u(\cdot), v(\cdot)$ are chosen by the players from the sets Ω_U, Ω_V. Let $K(t)$ be a matrix function satisfying following conditions:

1. $K(t) = 0$ for $t < 0$,
2. $K(0) = E$,
3. $K(t)$ is continuous on $[0, +\infty)$,
4. $K(t)$ satisfies the equation $K(t) = AK(t) + BK(t-\tau)$.

By Cauchy formula, the trajectory of (2.58), under given controls of the players, has a form (Bellman and Cooke 1963):

$$z(t) = K(t)z(0) + \int_{-\tau}^{0} K(t-\tau-s)Bz^0(s)ds - \int_{0}^{t} K(t-s)\phi_2(u(s), v(s))ds, t > 0.$$

Let us apply the general scheme presented in the Section 2.4. In the studied case $t_0 = 0$,

$$g(t) = K(t)z(0) + \int_{-\tau}^{0} K(t-\tau-s)Bz^0(s)ds, \ \phi(t,s,u,v) = K(t-s)\phi(u,v).$$

For $0 \le t \le T \le +\infty$

$$z(T,t,u_t(\cdot),v_t(\cdot)) = K(T)z(0) + \int_{-\tau}^{0} K(T-\tau-s)Bz^0(s)ds$$
$$-\int_{0}^{t} K(T-s)\phi(u(s),v(s))ds,$$
$$z(T) = z(T,t,u_t(\cdot),v_t(\cdot)) - \int_{t}^{T} K(T-s)\phi(u(s),v(s))ds,$$

It is evident that $z(T) \in M^*$, if and only if $\pi z(T) \in M$, π – an orthoprojector from R^n onto L. In its turn, $\pi z(T) \in M$ if and only if for any admissible control $v(s)$, $s \in [t, T]$, there exists an admissible control $u(s)$, $s \in [t, T]$, such that:

$$\pi z(T,t,u_t(\cdot),v_t(\cdot)) \in \bigcap_{v(\cdot)\in\Omega_V} \left[M + \int_{t}^{T} \pi K(T-\theta)\, \phi(U, v(\theta))d\theta \right]. \tag{2.59}$$

Let us denote

$$W(T,t)=\bigcap_{v(\cdot)\in\Omega_V}\left[M+\int_t^T \pi K(T-\theta)\,\phi_2(U,v(\theta))\,d\theta\right].$$

It is clear that at $t_0=0\ W(T)=W(T,0)$ (2.8). By the time of "first absorption" we call a moment:

$$T_0=\min\left\{T:T\geq 0,\pi\, K(T)z(0)+\int_{-\tau}^{0}\pi\, K(T-\tau-s)Bz^0(s)ds\in W(T,0)\right\}$$
$$=\min\{T:T\geq 0,\pi\, z(T,0,u_0(\cdot),\ v_0(\cdot))\in W(T,0)\}.$$

On the other hand, for $\zeta=\zeta(t)=z(T,t,,u_t(\cdot),v_t(\cdot))$:

$$W(T,t,\zeta;p)=(\zeta,p)+\int_t^T \min_{v\in V}\max_{u\in U}\left(K(T-s)\,\phi(u(s),v(s)),-p\right)ds,$$
$$\lambda(T,t,\zeta)=\min_{\|p\|=1,\,p\in L}\left[W(T,t,\zeta;p)+C(M;-p)\right],$$
$$T_0=\min\{T:T\geq 0,\lambda(T,0,g(T))\geq 0\},$$
$$\zeta(0)=z(T,0,u_0(\cdot),v_0(\cdot))=g(T)=K(T)z(0)+\int_{-\tau}^{0}K(T-\tau-s)Bz^0(s)ds.$$

Let us recall the notations of the general scheme:

$$T_t=T(t,,u_t(\cdot),v_t(\cdot))=\min\{T;T\geq t,\lambda(T,t,z(T,t,u_t(\cdot),v_t(\cdot)))\geq 0\},$$
$$\Gamma(t,T,\zeta)=\{p:\|p\|=1,p\in L,W(T,t,\zeta;p)+C(M;-p)=\lambda(t,T,\zeta)\}.$$

As before, the pair $(t,z(T_t,t,u_t(\cdot),v_t(\cdot)))$ stands for a position at the time $t,t\geq 0$. The initial position is the pair:

$$\left(0,K(T)z(0)+\int_{-\tau}^{0}K(T-\tau-s)Bz^0(s)ds\right).$$

The analog of the theorem 2.1 as applied to the differential-difference game (2.58) is presented below.

Theorem 2.2. *Let in the linear differential-difference game* (2.58) *following conditions be satisfied:*

1. $W(T,t)\neq\varnothing$ *for* $0\leq t\leq T\leq T_0, T_0<+\infty$;
2. *for any position* $(t^*,z(T_{t^*},t^*,u_{t^*}(\cdot),v_{t^*}(\cdot)))$, $0\leq t^*\leq T_0$, *such that for* $T_{t^*},T_{t^*}\leq T_0$, *a neighborhood* $\Omega(t^*,z(T_{t^*},t^*,u_{t^*}(\cdot),v_{t^*}(\cdot)))$ *of this point exists, such that for all* (t,ζ) *from this neighborhood the set-valued mapping* $\Gamma(T_{t^*},t,\zeta)$ *consists of a unique element. Then the game can be terminated no later than at the time of "first absorption"* T_0.

The proof is analogous to that of theorem 2.1.

2.9 TIME OF THE "FIRST ABSORPTION" IN THE LINEAR NONSTATIONARY DIFFERENTIAL GAME

Let motion of the conflict-controlled process be described by the equation:

$$\dot{z}(t) = A(t)z + B(t)u + C(t)v, t \geq t_0, z(t_0) = z_0. \tag{2.60}$$

Here $A(t), B(t), C(t)$ are matrices of dimensions of $n \times n, n \times r, n \times s$, respectively with continuous functions as their elements. Parameters u and v are controls, chosen by the pursuer and the evader from the sets U, $V, U \in coK(R^n), V \in coK(R^n)$, $u(t)$ and $v(t)$ are measurable functions, $t \geq t_0$.

We consider the pursuit game with complete information, namely, information on current state $z(t)$ of the system (2.60) is available to the players at each instant of time $t, t \in [t_0, \infty)$. The pursuer strives in the shortest time to bring a trajectory of the system (2.60) to a given terminal set $M(t)$, that is to achieve inclusion $z(t) \in M(t)$, where $M(t)$, $M(t) \neq \varnothing$, $t \in [t_0, \infty)$, is a closed convex-valued mapping. The goal of the evader is opposite.

Suppose that at each $t \in [t_0, \infty)$ the support function of the set-valued mapping $M(t)$:

$$W(M(t)\ ; p) = \sup_{x \in M(t)} (x, p),$$

is continuous in p on the barrier cone:

$$K_{M(t)} = \{p : p \in R^n, W(M(t); p) < +\infty\}.$$

Note that this cone is closed for all $t \in [t_0, \infty)$.

In what follows, the pair $(t, z(t))$ will be referred to as the game position at the moment t. It is said that the game of pursuit (2.60) with the terminal set $M(t)$ can be terminated no later than at the time $T(t_0, z_0)$, beginning from the initial position (t_0, z_0), if the pursuer by the way of the control choice, on the basis of current information on system state, can bring a trajectory of the object to the terminal set at certain moment of time $t, t_1 \leq T(t_0, z_0)$, that is $z(t_1) \in M(t_1)$, under any admissible control of the evader $v(t), t \in [0, t_1)$.

Given the initial position (t_0, z_0) and the controls $u_t(\cdot)$ and $v_t(\cdot)$, by the Cauchy formula, solution to the system (2.60) has the form

$$z(t) = \Phi(t, t_0) z_0(t) + \int_{t_0}^{t} \Phi(t, \theta)\, B(\theta) u(\theta) d\theta + \int_{t_0}^{t} \Phi(t, \theta)\, C(\theta) v(\theta) d\theta,$$

where $\Phi(t, \theta)$ is the transition (Cauchy) matrix of homogeneous system $\dot{z} = A(t)z$.

Note that this matrix has semi-group property. In view of this fact, for $t_0 \leq t \leq T < +\infty$, we set $z = z(t)$,

$$W(T, t, z, p) = (\Phi^*(T, t) p, z) + \int_{t}^{T} \max_{u \in U} (\Phi^*(T, \theta) p, B(\theta) u) d\theta$$
$$+ \int_{t}^{T} \min_{v \in V} (\Phi^*(T, \theta) p, C(\theta) v)\, d\theta,$$
$$\lambda(T, t, z) = \min_{\|p\|=1} \left[W(T, t, z, p) + W(M(T); -p)\right],$$
$$\Gamma(T, t, z) = \{\, p : \|p\| = 1, W(T, t, z, p) + W(M(T); -p) = \lambda(T, t, z)\}.$$

Here $\Phi^*(T, \theta)$ – matrix function which is conjugate to the matrix $\Gamma(T, \theta)$, $p \in R^n, \Phi(T, t, z) \subset -K_{M(T)}$. If $z(T) \notin M(T)$, then using consideration similar to that of section 2.4.

$$\lambda(T, t, z) = \min_{\|p\|=1} \left[W(T, t, z; p) + C(M(t), -p)\right] < 0.$$

Let us denote by $T(t,z)$ the least root of the equation in T: $\lambda(T,t,z)=0, 0\le t\le T<+\infty$. In case there no such root we set $T(t,z)=+\infty$. Evidently, $T(t,z)=t$ only when $z(t)\in M(t)$. The time $T(t_0,z_0)$ will be referred to as the time of "first absorption".

Theorem 2.3. *Let for the position* (t_1,z_1), *such that* $T(t_1,z_1)<+\infty$,

set $\Gamma(T(t_1,z_1),t,z)$ consists of a unique vector $p(T(t_1,z_1),t,z)$ for all points (t,z) from some neighborhood of (t_1,z_1);

$\max_{u\in U}(\Phi^*(T(t_1,z_1),t)p(T(t_1,z_1,),t,z),B(t)u)$ is attained on a unique vector $u(T(t_1,z_1),t,z)$ for all points of this neighborhood.

Then the pursuer can terminate the game no later than at the time $T(t_0,z_0)$.

Proof. Suppose that for the given initial position (t_0,z_0) the time $T_0=T(t_0,z_0)$ is evaluated. Denote by $\Omega(t_0,z_0)$ the neighborhood of (t_0,z_0), in which, vector $p(T_0,t,z)$ is uniquely defined, by assumptions of the theorem. Since function $W(T_0,t,z,p)+W(M(T_0),-p)$ is jointly continuous in variables t,z,p then function $p(T_0,t,z)$ is jointly continuous in t, z within $\Omega(t_0,z_0)$. By assumption b), with help of the equality:

$$(\Phi^*(T_0,t)\ p(T_0,t,z),\ B(t)\,u^0(T_0,t,z))=\max_{u\in U}(\Phi^*(T_0,t)\ p(T_0,t,z),\ B(t)\,u) \qquad (2.61)$$

the control $u^0(T_0,t,z)$ defined in a unique way. That is why, function $u^0(T_0,t,z)$ is jointly continuous in t, z.

Let us set the pursuer's control within $\Omega(t_0,z_0)$ to be equal $u^0(T_0,t,z)$. Consider the following equation:

$$\dot z(t)=A(t)z+B(t)u^0(T_0,t,z)+C(t)\,v(t),\quad z(t_0)=z_0.$$

For any admissible evader's control $v(t)$ there exists a solution to this equation which can be extended to the boundary of $\Omega(t_0,z_0)$. Suppose that it reaches the boundary of $\Omega(t_0,z_0)$ at the time t_1. If $z^0(t_1)\in M(t_1)$ then the game is terminated. If, otherwise, $z^0(t_1)\notin M(t_1)$ then $\lambda(t_1,t_1,z^0(t_1))<0$.

Thus, a trajectory of system (2.60) is constructed on some interval $[t_0,t_1]$. Let us prove the inequality:

$$T(t_1,z^0(t_1))\le T(t_0,z_0). \qquad (2.62)$$

To this end, it suffices to show that:

$$\lambda(T(t_0,z_0),t,z^0(t))\ge 0,\quad t_0\le t\le t_1.$$

Let us find the derivative in t of function $\lambda(T_0,t,z^0(t))$ with account of the pursuer's control chosen on the interval $[t_0,t_1]$. As the set $\Gamma(T_0,t,z)$ consists of a unique vector $p(T_0,t,z)$ within the set $\Omega(t_0,z_0)$, then function $\lambda(T_0,t,z)$ has continuous partial derivatives in t and z there. Now we evaluate these derivatives.

$$\frac{\partial}{\partial t}\lambda(T_0,t,z^0(t))=-(A^*(t)\,p(T_0,t,z^0(t)),\Phi(T_0,t)z^0(t))$$
$$-\max_{u\in U}(\Phi^*(T_0,t)\,p(T_0,t,z^0(t)),\ B(t)u)-\min_{v\in V}(\Phi^*(T_0,t)\,p(T_0,t,z^0(t)),\ C(t)v),$$
$$\frac{\partial}{\partial z}\lambda(t,z^0(t),T_0)=(\Phi^*(T_0,t)\,p(T_0,t,z^0(t)),A(t)z^0(t)+B(t)u^0(T_0,t,z^0(t))+C(t)\,v(t)).$$

Then,

$$\frac{d\lambda(T_0,t,z^0(t))}{dt} = -\left(A^*(t)\,p(T_0,t,z^0(t)),\, \Phi(T_0,t)z^0(t)\right)$$
$$-\max_{u\in U}\left(\Phi^*(T_0,t)\,p(T_0,t,z^0(t)),\, B(t)u\right) - \min_{v\in V}\left(\Phi^*(T_0,t)\,p(T_0,t,z^0(t)),\, C(t)v\right)$$
$$+\left(\Phi^*(T_0,t)\,p(T_0,t,z^0(t)),\, A(t)z^0(t) + B(t)u^0(T_0,t,z^0(t)) + C(t)\,v(t)\right).$$

In view of the control choice (2.61) from the above formula there follows that $\lambda(T_0,t,z^0(t)) \geq 0$ on the interval $[t_0,t_1]$, that is function $T(t,z^0(t))$ does not decrease along the trajectory $z^0(t), t\in[t_0,t_1]$. Since $\lambda(T_0,t_0,z_0)=0$, then $\lambda(T_0,t_1,z^0(t_1))\geq 0$. Then, in view of the fact that $T(t_1,z^0(t_1))$ is minimal $T, t\geq t_0$, such that $\lambda(t_1,z^0(t_1),T)=0$, then the inequality (2.62) is valid. Hence, the trajectory of system (2.60) is built on certain interval $[t_0,t_1]$. Note that the constructed control $u^0(t), t\in[0,t_1)$, appears as piecewise continuous function.

Analogously, beginning from the position $(t_1,z^0(t_1))$ a solution of the equation (2.60) on some interval of time $[t_1,t_2]$ can be constructed a such way, that the following inequality be held:

$$T(t_2,z^0(t_2)) \leq T(t_1,z^0(t_1)).$$

By preceding to perform this procedure over and over, we obtain a sequence of inequalities:

$$T(t_k,z^0(t_k)) \leq T(t_{k-1},z^0(t_{k-1})) \leq \ldots \leq T(t_2,z^0(t_2)) \leq T(t_1,z^0(t_1)),\, t_{k-1}\leq t_k, \quad k=1,2,\ldots$$

It may occur that $T(t,z^0(t))=t$ for some $t\in[t_{k-1},t_k), t_{k-1}\leq t_k$. That would mean termination of the game since $T(t,z(t))=t$ if and only if $z(t)\in M(t)$. Otherwise, the process of control construction can be continued. Because of the double inequality to hold:

$$t \leq T(t,z^0(t)) \leq T_0,$$

the equality $T(t,z^0(t))=t$, pointing to the game termination, will be satisfied no later than at the time T_0.

As an illustration, we refer to the non-stationary variant of Pontryagin's Testing Example. Let motions of the pursuer and the evader be described by the systems of differential equations:

$$\dot z_1 = z_2,$$
$$\dot z_2 = -\alpha(t)z_2 + \rho u\,,$$

and

$$\dot z_3 = z_4,$$
$$\dot z_4 = -\beta(t)z_4 + \sigma v,$$

respectively, where $z_i \in R^n, i=1,\ldots,4, n\geq 2; U=\{u:\|u\|\leq 1\}; V=\{v:\|v\|\leq 1\}$. The terminal set is given in the terms of variable z, $z=(z_1,z_2,z_3,z_4)$:

$$M(t)=\{z:\|z_1-z_3\|\leq \varepsilon(t)\}.$$

It is supposed that functions $\alpha(t),\beta(t),\varepsilon(t), t\in[t_0,+\infty)$ are continuous, $\varepsilon(t)\geq 0$, $0<\alpha_1\leq\alpha(t)\leq\alpha_2, 0<\beta_1\leq\beta(t)\leq\beta_2,\ \alpha(t)\neq\beta(t)$.

Introducing the notation:

$$p=(p_1,p_2,p_3,p_4),\ p_i\in R^n,\ i=\overline{1,4},$$

we deduce the formulas:

$$W(M(t),p)=\begin{cases}\|p_1\|\varepsilon(t), & p_1=-p_3\\ +\infty, & p_1\neq -p_3\end{cases},$$

$$K_{M(t)}=\{p: p=(p_1,0,-p_1,0), p_1\in R^n\},$$

$$W(M(t),p)=\|p_1\|\varepsilon(t),\ p\in K_{M(t)},$$

$$\lambda(T,t,z)=\min_{\|p_1\|=1}\left[\left(p_1,z_1-z_3+l_1(T,t)z_2-f_1(T,t)z_4\right)+\rho\int_t^T l_1(T,r)dr-\sigma\int_t^T f_1(T,r)dr+\frac{1}{\sqrt{2}}\varepsilon(T)\right],$$

where:

$$l_1(T,t)=\int_t^T l_0(\theta,t)d\theta,\quad l_0(\theta,t)=\exp\left(-\int_t^\theta \alpha(r)dr\right),$$

$$f_1(T,t)=\int_t^T f_0(\theta,t)d\theta,\quad f_0(\theta,t)=\exp\left(-\int_t^\theta \beta(r)dr\right).$$

It should be noted, that the minimum in p ($\|p\|=1$), in the expression for $\lambda(T,t,z)$ is attained at the vector $p(T,t,z)=(p_1(T,t,z),0,-p_1(T,t,z),0)$, where:

$$p_1(T,t,z)=-\frac{1}{\sqrt{2}}\frac{z_1-z_3+l_1(T,t)z_2-f_1(T,t)z_4}{\|z_1-z_3+l_1(T,t)z_2-f_1(T,t)z_4\|},$$

under the condition that the denominator in the above formula for $p_1(T,t,z)$ does not turn into zero. Function $T(t,z)$ is defined by the following equality in T:

$$\|z_1-z_3+l_1(T,t)z_2-f_1(T,t)z_4\|=\phi(T,t)+\frac{1}{\sqrt{2}}\varepsilon(T), \tag{2.63}$$

where:

$$\phi(T,t)=\rho\int_t^T l_1(T,r)dr-\sigma\int_t^T f_1(T,r)dr.$$

It is easy to see that the conditions of the theorem are fulfilled if $\phi(T,t)>0$ for all $T\in[t_0,+\infty)$, $T>t$, that, in the turn, has place if:

$$\rho l_1(T,t)-\sigma f_1(T,t)>0. \tag{2.64}$$

Denoting $g(T,t)=\dfrac{l_1(T,t)}{f_1(T,t)}$ we now show that function $g(T,t)$ take values more or equal $\min\left\{1,\beta/\alpha\right\}$. In view of the constraints, imposed on the $\alpha(t)$, $\beta(t)$, we infer:

$$\frac{\beta_1}{\alpha_2}\frac{1-e^{-\alpha_2(T-t)}}{1-e^{-\beta_1(T-t)}}\le g(T,t)\le\frac{\beta_2}{\alpha_1}\frac{1-e^{-\alpha_1(T-t)}}{1-e^{-\beta_2(T-t)}}. \tag{2.65}$$

We will separately study the cases $\beta_1 > \alpha_2$ and $\beta_1 \le \alpha_2$.

Let us suppose that $\beta_1 > \alpha_2$. It is evident that $\lim_{t \to T} g(T,t) = 1$. We now show that if $T \in [t_0, +\infty)$, $T > t$, then $g(T,t) \ne 1$. Suppose the opposite, namely, let for some fixed $t_1 \in [t_0, +\infty)$ there exist $T_1 \in [t_1, +\infty)$, such that $g(T_1, t_1) = 1$. This means that at some moment of time $T, T \in [t_1, T_1)$,

$$\frac{d}{dt}\left(l_1(T,t_1) - f_1(T,t_1)\right) = l_0(T,t_1) - f_0(T,t_1) = 0.$$

As $\alpha(r) \ne \beta(r)$ for all $r \in [t_0, +\infty)$, it is impossible, so we come to contradiction. It remains to prove that $g(T,t) > 1$. In fact, from the inequalities (2.65) there follows that:

$$\lim_{T \to +\infty} g(T,t) = \frac{\beta_1}{\alpha_2} > 1.$$

We see that the inequality (2.64) is fulfilled if:

$$\beta_1 > \alpha_2, \quad \rho \ge \sigma. \tag{2.66}$$

If, otherwise, $\beta_1 \le \alpha_2$, then $\dfrac{1 - e^{-\alpha_2 (T-t)}}{1 - e^{-\beta_1 (T-t)}} \ge 1$, when taking into account (2.65) we obtain inequality $g(s,t) \ge \dfrac{\beta_1}{\alpha_2}$, in which the equality can occur only if $\beta_1 = \alpha_2$. Therefore, in the case when:

$$\beta_1 \le \alpha_2, \quad \frac{\rho}{\alpha_2} \ge \frac{\sigma}{\beta_1}, \tag{2.67}$$

in which at least one of the inequalities is strict, the relationship (2.64) is satisfied.

Thus, if one of the inequalities (2.66) or (2.67) are fulfilled, then the conditions of the theorem are met. In this case the control of the pursuer, realizing the pursuer's goal is given by the formula:

$$u(T,t,z) = \sqrt{2}\, p_1(T,t,z).$$

Hence, the game can be terminated beginning from all positions $(t, z(t))$, for which the equation (2.63) has solution $T(t,z)$ no less than t, no later than at the time $T(t,z)$. In case of strict inequalities in (2.66), (2.67), that is, when a constant k exists, such that:

$$\frac{\rho}{\alpha_2} - \frac{\sigma}{\beta_1} > k \text{ if } \beta_1 \le \alpha_2 \text{ or } \rho - \sigma > k \text{ if } \beta_1 > \alpha_2,$$

Function $\phi(s,t)$ increases no slower than a linear function with coefficient k.

At the beginning of the game $\|z_1^0 - z_3^0\| > \varepsilon(t_0)$ as T grows, the left-hand side of (2.63) remains bounded, that is why the equation (2.63) has a finite solution. Then, as follows from the above considerations, the game of pursuit can be terminated in a finite period from any initial position (t_0, z_0). This result agrees with the conditions obtained in the stationary case (Pshenichnyi 1968).

2.10 CONFLICT-CONTROLLED IMPULSE SYSTEMS

The impulse processes, in their number those affected by pushes and thrusts, are thoroughly studied in the monographs (Samoilenko and Perestyuk 1987, Krivonos et al. 2005,

Zavalishchin and Sesekin 1991), containing in addition, extensive bibliography on this subject. Specific feature of the impulse process is that its solution appears as discontinuous function, more precisely, as function of bounded variation without singular component.

By the Lebesgue theorem (Natanson 1957), such solution is a sum of absolutely continuous function and function of no more than countable number of jumps. Game problems for such processes are analyzed in the book (Krivonos et al. 2005), where the push-affected systems are explored on the basis of the method of resolving functions (Chikrii A.A. 1997, Chikrii A.A. 2010, Chikrii A.A. and Chikrii G.Ts. 2014).

In this section we employ a solution representation, going back to the monograph (Samoilenko and Perestyuk 1987), in which the explicit form of the Dirac δ- function is not used. Let us consider a quasilinear conflict-controlled process:

$$\dot{z} = A(t)z + \phi(t,u,v),\ z \in R^n,\ t \geq t_0 \geq 0,\ z(t_0) = z_0, u \in U(t), v \in V(t). \tag{2.68}$$

Here $A(t)$ – is a matrix function with elements which are added together on arbitrary finite time interval, $U(t)$ and $V(t)$ (the players control domains) are measurable compact-valued mappings, function $\phi(t, u, v)$ (the control unit) satisfies the Caratheodori condition (jointly continuous in u, v and measurable in t). Besides, the following inequality is assumed to hold:

$$\|\phi(t,u,v)\| \leq c(t) \ \ \forall\, u \in U(t), v \in V(t), t_0 \leq t < +\infty, \tag{2.69}$$

where $c(t)$ is a scalar function, sum up on an arbitrary finite time interval.

The process (2.68) is impulse affected (Samoilenko and Perestyuk 1987), namely a trajectory of the system (2.68) has the first-order discontinuities at the given points:

$$\tau_i,\ \tau_1 < \tau_2 < \ldots < \tau_i < \tau_{i+1} < \ldots < +\infty,$$

the consequence of these points being free of accumulation points. The value of jump at the moment τ_i has the form:

$$\Delta z\Big|_{t=\tau_i} = z(\tau_i + 0) - z(\tau_i) = B_i z_i + a_i,\ i = 1,2,..., \tag{2.70}$$

where B_i are constant matrices, $z_i = z(\tau_i)$, a_i, $a_i \in R^n$ are vectors. Therefore, $z(t)$ is continuous from the left side till the jump. Also, a terminal set $M^*(t)$ is given, having cylindrical form:

$$M^*(t) = M_0 + M(t),\ t_0 \leq t < +\infty, \tag{2.71}$$

where M_0 is a linear subspace of R^n, $M(t)$ – a measurable compact-valued mapping taking its values in L – orthogonal complement to M_0 in R^n.

The goal of the first player (u) is, by acting at the process with help of admissible controls, constructed on the basis of information on current position (t, z), in the shortest time to drive a trajectory of the system (2.68) – (2.70) to the terminal set, under any counteraction of the second player (v).

Admissible controls $u(\tau)$ and $v(\tau)$ chosen, solution to the system (2.1) has the form (an analog of the Cauchy formula):

$$z(t) = \Phi(t,t_0)z_0 + \int_{t_0}^{t} \Phi(t,\tau)\ \phi(\tau,u(\tau),v(\tau))d\tau + \sum_{t_0<\tau_i<t} \Phi(t,\tau_i)a_i, \tag{2.72}$$

where:

$$\Phi(t,t_0)=\Omega(t,\tau_{j+k})(E+B_{j+k})\prod_{\nu=k}^{1}\Omega(\tau_{j+\nu},\tau_{j+\nu-1})\cdot(E+B_{j+\nu-1})\Omega(\tau_j,t_0),$$
$$\tau_{j-1}<t_0\le\tau_j<\tau_{j+k}<t\le\tau_{j+k+1},$$
$$\Phi(t,\tau)=\Omega(t,\tau_{j+k})\prod_{\nu=k}^{s+1}(E+B_{j+\nu})\Omega(\tau_{j+\nu},\tau_{j+\nu-1})(E+B_{j+s})\cdot\Omega(\tau_{j+s},\tau),$$
$$\tau_{j-1}<t_0\le\tau_j\le\tau_{j+s-1}<\tau\le\tau_{j+s}<\tau_{j+k}<t\le\tau_{j+k+1}.$$

Here $\Omega(t,\tau)$ is a matrizant of the homogeneous system $\dot{z}=A(t)z$ and E is a unit matrix.

It is readily seen that the solution representation (2.72) presents itself a particular case of (2.1). That is why the scheme of positional approaching the terminal set (see Sections 2.3 and 2.4) is applicable in the case under study.

2.11 GAME PROBLEMS FOR PROCESSES WITH FRACTIONAL DERIVATIVES

In this section we consider the conflict-controlled processes whose evolution is described by the fractional-order systems. Formulas for the solution representation are given pointing to the fact that positional approach is applicable to study these processes. In so doing, we restrict ourselves to the cases of classical derivatives of Riemann – Liouville and the regularized derivatives of Dzharbashyan – Nersesyan – Caputo. As follows from the results of works (Chikrii A.A., Matychyn I., Gromaszek K. and Smolarz A. 2011, Chikrii A.A. a 2008, Chikrii A.A. b 2008) similar representation can be obtained in the cases of the Miller – Ross and Hilfer sequential derivatives.

It should be mentioned that analogous game problems are studied in (Chikrii A.A., Matychyn I., Gromaszek K. and Smolarz A. 2011) on the basis of the method of resolving functions.

Let R^m be the m-dimensional Euclidean space, R_+ be the positive semi-axis and N – the set of natural numbers. Suppose that $f(t), f:R_+\to R^m$, is an absolutely continuous function.

Here we consider fractional derivatives of arbitrary order. Suppose that $n-1<\alpha<n, n\in N$. Let us denote the fractional part of α by $\{\alpha\}$ and its integer part by $[\alpha]$, namely:

$$[\alpha]=n-1,\{\alpha\}=\alpha-n+1.$$

The Riemann – Liouville fractional derivative of arbitrary order $\alpha\,(n-1<\alpha<n, n\in N)$ is defined as folows (Samko, Kilbas and Marichev 1993):

$$D^\alpha f(t)=\left(\frac{d}{dt}\right)^{[\alpha]}D^{\{\alpha\}}f(t)=\left(\frac{d}{dt}\right)^{[\alpha]}\frac{1}{\Gamma(1-\{\alpha\})}\frac{d}{dt}\int_0^t\frac{f(\tau)}{(t-\tau)^{\{\alpha\}}}d\tau=\frac{1}{\Gamma(n-\alpha)}\left(\frac{d}{dt}\right)^n\int_0^t\frac{f(\tau)}{(t-\tau)^{\alpha-n+1}}d\tau. \tag{2.73}$$

Lemma 2.3. Let $n-1<\alpha<n, n\in N$, *and let the function* $f(t)$ *have absolutely continuous derivatives up to the order* $n-1$. *Then the following formula is true:*

$$D^\alpha f(t)=\sum_{k=0}^{n-1}\frac{t^{k-\alpha}}{\Gamma(k-\alpha+1)}f^{(k)}(0)+\frac{1}{\Gamma(n-\alpha)}\int_0^t\frac{f^{(n)}(\tau)}{(t-\tau)^{\alpha-n+1}}d\tau. \tag{2.74}$$

As before, the integral term in (2.74) is the regularized Caputo derivative of order $\alpha(n-1<\alpha<n, n\in N)$:

$$D^{(\alpha)}f(t)=\frac{1}{\Gamma(n-\alpha)}\int_0^t\frac{f^{(n)}(\tau)}{(t-\tau)^{\alpha+1-n}}d\tau=D^\alpha f(t)-\sum_{k=0}^{n-1}\frac{t^{k-\alpha}}{\Gamma(k-\alpha+1)}f^{(k)}(0). \tag{2.75}$$

In (Chikrii A.A. 2008) the generalized Mittag-Leffler matrix function was introduced:

$$E_\rho(B;\mu)=\sum_{k=0}^{\infty}\frac{B^k}{\Gamma(k\rho^{-1}+\mu)},$$

where $\rho>0, \mu\in C$ and B is an arbitrary square matrix of order m.

Consider the conflict-controlled process, described by the equation with the classic Riemann – Liouville fractional derivative:

$$D^\alpha z = Az+\phi(u,v),\, z\in R^m,\, n-1<\alpha<n,\, u\in U,\, v\in V,\, U\in K(R^m),\, V\in K(R^m), \quad (2.76)$$

where A is a square matrix of order m and $\phi(u, v)$ – function, jointly continuous in its variables. The initial conditions are:

$$D^{\alpha-k}z(t)\big|_{t=0}=z_{k1}^0,\, k=1,2,\ldots,n. \quad (2.77)$$

Under the chosen players controls a trajectory of the system (2.76), (2.77) has the form (Chikrii A.A., Matychyn I., Gromaszek K. and Smolarz A. 2011):

$$z(t)=\sum_{k=1}^{n}t^{\alpha-k}E_{1/\alpha}\left(At^\alpha;\alpha-k+1\right)z_{k1}^0+\int_0^t(t-\tau)^{\alpha-1}E_{1/\alpha}\left(A(t-\tau)^\alpha;\alpha\right)\,\phi(u(\tau),v(\tau))d\tau. \quad (2.78)$$

In the case of the process with the regularized derivative (Chikrii A.A., Matychyn I., Gromaszek K. and Smolarz A. 2011):

$$D^{(\alpha)}z = Az+\phi(u,v),\, z\in R^m,\, n-1<\alpha<n,\, u\in U,\, v\in V, \quad (2.79)$$

and the Cauchy initial data:

$$z^{(k)}(0)=z_{k2}^0,\, k=0,1,\ldots,n-1, \quad (2.80)$$

a trajectory of the system (2.79), (2.80) has the form:

$$z(t)=\sum_{k=0}^{n-1}t^{k}E_{1/\alpha}\left(At^\alpha;k+1\right)z_{k2}^0+\int_0^t(t-\tau)^{\alpha-1}E_{1/\alpha}\left(A(t-\tau)^\alpha;\alpha\right)\,\phi(u(\tau),v(\tau))d\tau. \quad (2.81)$$

As the representations (2.78), (2.81) are of the form (2.1), the sufficient conditions for positional approaching the terminal set in the time of "first absorption" can be derived in the both cases in accordance with the general scheme.

REFERENCES

Aubin J.-P., Frankowska H. 1990: Set-valued analysis, Birkhauser, Boston.

Aumann R.J. 1965: Integrals of set-valued functions, J. Math. Anal. Appl., 12: pp. 1–12.

Bellman R., Cooke K.L. 1963: Differential-difference equations, Academic Press, New York-London.

Chikrii A.A. 1979: Quasilinear Problem of Approach with Many Participants, J. Appl. Maths. Mechs., 43(3), pp.486–491.

Chikrii A.A. 1997: Conflict-controlled processes, Boston; London; Dordrecht; Springer, 424 p.

Chikrii A.A. 2010: An analytic method in dynamic games, Trudy Mat. Inst. RAN im B.A. Steklova, 271, pp. 76–92.

Chikrii A.A. a 2008: Optimization of game interaction of fractional-order controlled systems, Optimization Methods and Software, 23(1), pp. 39–73.
Chikrii A.A. b 2008: Game dynamic problems for systems with fractional derivatives, Pareto Optimality, Game Theory and Equilibria, Springer, vol. 17, pp. 349–386.
Chikrii A.A., Chikrii G.Ts. 2014: Matrix resolving functions in dynamic games of approach, Cybernetics and Systems Analysis, 50(2), pp. 201–217.
Chikrii A.A., Matychyn I, Gromaszek K., Smolarz A. 2011: Control of fractional-order dynamic systems under uncertainty, Modelling and Optimization, Politechnika Lubelska, pp. 3–56.
Chikrii A.A., Rappoport J.S. 1978: Linear Problem of Pursuit by several Cuided Objects, Cybernetics, Plenum Publish. Corp., New York, 14 (3), pp. 407–415.
Chikrii G.Ts. 1995: An approach to the solution of linear differential games with variable information delay, Journal of Automation and Information Sciences: Beggel House, Inc. (USA), 27 (3&4): pp.163–170.
Chikrii G.Ts., Volyanskiy K.Yu. 2002: Position control in integro-differential pursuit games, Cybernetics and Systems Analysis, 38(5), pp.720–735.
Gabrielyan M.S., Kryazhymskii A.V. 1983: Differential game of approach-evasion with m terminal sets, Dokl. Acad. Nauk SSSR, 283(3), pp. 525–527.
Gaishun I.V. 1999: Introduction to the theory of linear nonstationary systems, Mathematics Institute of NAN of Belorussia, (in Russian).
Gantmakher F.R. 1987: Theory of matrices, American Mathematical Soc. Chelsea Publishing.
Ioffe A.D., Tikhomirov V.M. 1979: Theory of extremal problems, vol.6 of Studies in Mathematics and its Applications. North-Holland, Amsterdam.
Krasovskii N.N. 1970: Game Problems on the Encounter of Motions, Nauka, Moscow, (in Russian).
Krasovskii N.N. 1985: Control of dynamic system, Nauka, Moscow, (in Russian).
Krasovskii N.N., Subbotin A.I. 1974: Positional Differential Games, Nauka, Moscow, (in Russian).
Krivonos Yu.G., Matychyn I.I., Chikrii A.A. 2005: Dynamic games with discontinuous trajectories, Naukova Dumka, Kyiv, (in Russian).
Matychyn I, Chikrii A.A., Gromaszek K. 2012: Dynamics games involving impulses, Current problems in information and computational technologies, vol. 2, Politechnika Lubelska, pp. 51–106.
Mikhlin, S.G. 1960: Linear Integral Equations, Hindustan Publ. Corp., Delhi.
Natanson I.P. 1957: Theory of functions of real variable, Gos. Izd-vo of Techn-Theoretical Literature, Moscow, (in Russian).
Perekatov A.Eu., Chikrii A.A. 1993: Successive pursuit by position, Avtomatika I Telemekhanika, №. 10, pp. 86–95.
Pontryagin L.S. 1988: Selected scientific papers, Nauka, Moscow, 2, (in Russian).
Pshenichnyi B.N. 1968: Linear Differential Games, Avtomatika i Telemekhanika, №. 1, pp. 65–78, (in Russian).
Pshenichnyi B.N. 1969: Structure of differential games, Dokl. Acad. Nauk SSSR, 184(2), pp. 285–287, (in Russian).
Pshenichnyi B.N., Onopchuk Yu.N. 1968: Linear differential games under integral constraints, Izvestia AN SSSR, ser. Technicheskaya kibernetika, №. 1: pp.13–22, (in Russian).
Samko S.G., Kilbas A.A., Marichev O.I. 1993: Fractional Integrals and Derivatives, Gordon ang Breach, Amsterdam.
Samoilenko A.M, Perestyuk M.A. 1987: Differential equations with impulse effect, Vuscha Shkola, Kyiv, (in Russian).
Smirmov V.I. 1964: A Course of Higher Mathematics, Vol. IV: Integral Equations and Partial Differential Equations, Pergamon Press; First Edition edition.
Tarlinskii S.J. 1976: On a Certain Linear Differential Game of Convergence of Some Controlled Objects, Dokl. Akad. Nauk SSSR, 230(3), pp.853–857, (in Russian).
Zavalishchin S.T., Sesekin A.N. 1991: Impulse processes. Models and applications, Nauka, Moscow, (in Russian).

CHAPTER 3

Information technology for automated translation from inflected languages to sign language

Iu. Krak, O. Barmak & S. Romanyshyn
V.M. Glushkov Institute of Cybernetic of NAS of Ukraine, Kyiv, Ukraine

A. Kotyra & K. Gromaszek
Lublin University of Technology, Lublin, Poland

S. Smailova
D. Serikbayev East Kazakhstan State Technical University, Ust-Kamenogorsk, Kazakhstan

INTRODUCTION

This chapter describes the information technologies, that allow inflected languages to the sign languages (for hearing impaired/deaf people) for, using machine translation (Barmak et al. 2012), (Barmak et al. 2013), (Barmak et al. 2014), (Krak & Romanyshyn 2014), (Krak et al. 2011), (Krak et al. 2012), (Krak et al. 2014). The main aim of the research is to elaborate new computer systems for training and communication for the deaf people.

According to World Health Organization, there are about 70 million deaf people (and people with severe hearing loss) who use sign language as their first language or mother tongue. That are tens of millions of people for whom it is necessary create equivalent means for communication in society. Due to Paragraph 7 of the Rules Appendix 5 to UN Resolution 48/96 "Standard Rules on the Equalization of Opportunities for Persons with Disabilities" these needs have been expressed as follows: "We need to take care of that sign language was used for teaching deaf children, in their families and communities. We should also provide services in sign language to facilitate communication with other deaf people." If we consider that people with hearing disabilities need to communicate with family, friends, neighbors, the circle of people with whom they maintain contact will be hundreds of millions of people. Despite the rehabilitation work on verbal language learning, the communication with the deaf person is characterized by certain difficulties.

The main form of communication of deaf people is sign language. Sign language has national characteristics. People with hearing impairments use two sign languages which have different grammar and a different set of gestures for communication:

1. conversational sign language used in everyday communication and has its own (different from natural spoken) grammar;
2. tracing sign language used in official and business environment (includes gestures and manual alphabet, which is used to play words with the letters given), has no own grammar, uses the rules of national language grammar.

The rapid development of computer technology, its availability in a variety of mobile devices, development of appropriate technologies as software and hardware (distributed databases, cloud computing, Internet technologies, etc.) encourages their involvement in intensive solution of socially significant problems. One of these problems is the inclusion of deaf in active social life. The main obstacle to solving this problem is difficulty in communication between deaf and hearing people. Information technology for nonverbal communication of deaf could help to solve this problem.

Sign Language (SL) of deaf is a natural language that conveys information through movement of hands and fingers, facial expressions, position the body. It is used as a component for communication between people and serves as the primary means of communication for deaf people. Sign languages are not like ordinary visual interpretation. They have their own grammar, which can be used to discuss a variety of topics from simple and specific – to the sublime or abstract. Vocabulary of SL is still poorly understood, especially idiomatic, phraseology, morphology.

Automatic attribution of SL to all forms of verbal and written language is not entirely true, because you cannot consider SL in traditional models of linguistics. The most suitable description of the conversational SL morphology is the principle of value to the form. The values of "plurality", "time", "accessories", "expiry" in SL are expressed in specific ways. The way of doing gestures (quality of movement) changes depending on the importance of gesture. Due the problem of building automated translation from text to SL can be represented as forming the pairs of mutually unambiguous structures in inflected language and SL, that convey the sense of information. Flexion (ending) is used to express grammatical meanings in inflected languages (Indo-European and Semitic-hatski).

3.1 THEORETICAL PRINCIPLES OF AUTOMATED TRANSLATION SYSTEM DESIGN

Automatic translation of inflected languages into natural SL involves the establishment of appropriate information technology. The part of the technology is to apply specific conversion to input sentence in inflectional language that would give the original sentence in SL. These changes include the following mechanisms:

1. building correlations between gestures and words;
2. building correlations between grouped sentences (certain grammatical structures) in SL and inflected languages to ensure proper transmission of meaning;
3. expanding the set of existing correlations (between words and gestures, grammatical structures) to improve the quality of translation;
4. translation prediction in cases where the sentence didn't match existing grammatical structures.

Transfer approach was selected to build the automated translation system (Dugast et al. 2007) (Senellart 2006). The basic idea of building an automated translation system is to use bilingual corps of inflectional language sentences and corresponding sentences in SL. Bilingual pairs of sentences are analyzed on this base. For each word or the inflectional language sentence the corresponding gesture of SL sentence is determined. Similarly, in structure sentences are grouped as grammar translation constructions. Grammar translation constructions have set of words (that can be used in fixed positions) instead of words. After analyzing the grammar translation constructions based on the inflectional parameters of the set of words, the sets are summarized and can be represented as generalized grammar constructions of translation. Generalized grammar constructions allow to determine the grammatical structure of sentences corresponding in SL by analyzing the inflectional parameters of input sentence.

Chart of the information technology area

The goal of characterized translation information technology is to implement it as a Web service, that will use the software on different platforms, including:

1. automated translation of inflected language sentences to SL sentences,
2. provide functionality to extend the existing system,
3. statistically process the translated sentences for self-learning.

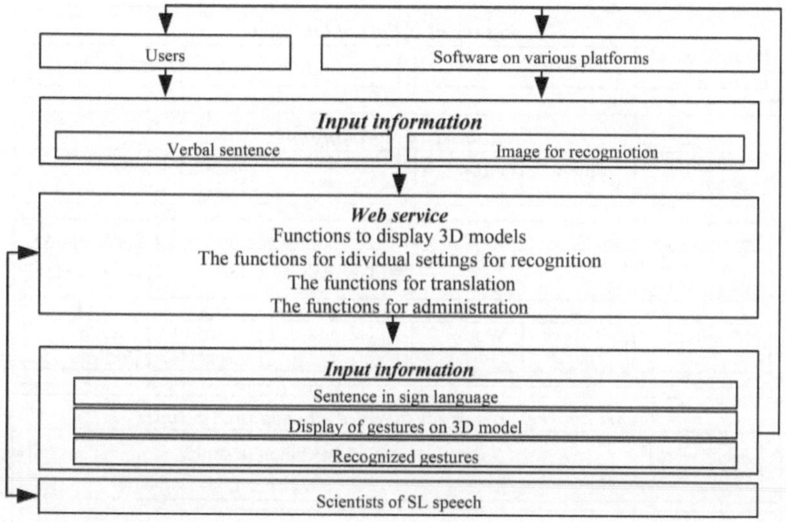

Figure 3.1. Block diagram of the information technology for automated translation system from inflected languages to sign language.

The term "Web service" describes a standardized way of integrating web-based, Internet applications using common standards and network protocols. Web service is a software application or information resource, access to which is carried out remotely using standard Internet protocols. It is important, that Web services enable access to software components, regardless of how these components work, where and in what platform they are located, and who offered physical or informational resources. Web services support the integration and interaction with other software components using XML-like messages. Interfaces and communication Web services are defined, described and are also using the XML language.

To achieve this goal, the information technology shown on the block diagram (Fig. 3.1) is proposed. Its essence lies in the following steps of transforming the input:

1. Users submit input information for translation. Entry information can be presented in the form of text, image recognition or audio file. Using software on various platforms, information is transmitted to a processing web service;
2. Web service should contain functions to transfer the input information from inflectional language to sign language; functions to display the translated information as 3D-models using individual settings of recognition of sign language; functions for system administration and extension;
3. Translated sign language sentence is returned to the user. When automatic translations of input sentence cannot be performed, system tries to forecast translation and allows user to choose the most adequate. The translation may be provided in the form of data information on 3D models, or descriptions of gestures in text format;
4. The scientists of SL have access to the service with the possibility of the administration for handling unusual situations and correction of the system.

To perform translation from inflected language to sign language the set models of inflected language are used (Shirokov 1998), (Shirokov 2004), (Shirokov et al. 2005) to enable appropriate modeling of related inflected and sign language grammar. Only simple sentences translation will be considered. Subject and predicate of sentences are linked with predicate bond. Any complex sentence can be represented as a combination of the simple sentences.

The presence of the following components is necessary for the operation of automated translation (see. Fig. 3.2):

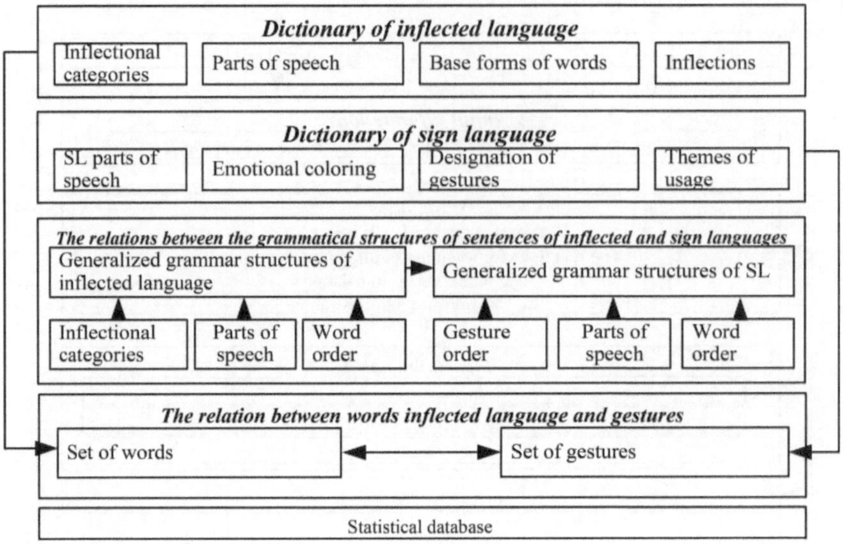

Figure 3.2. Block diagram of computer-aided translation components.

1. dictionaries of inflected and sign languages;
2. relations between grouped sentences (grammatical structures) of inflected and sign languages;
3. relations between words of inflected language and gestures;
4. statistical database for self-learning;
5. In order to support this solution, the dictionary of inflected language with the following contents;
6. the set of basic word forms (basic word forms are used as base for all the other word forms, such as a for noun in the Ukrainian language the basic word form is nominative singular, for verb—infinitive);
7. the set of inflections (by flexion we define how basic word forms are converted to all other word forms: endings, affixes, postfixes, changes of extension, changes in the basis of word);
8. the set of grammatical categories of inflection (data sets can contain number, gender, case, person, time, status, mode depending on the morphological features of language);
9. the set of parts of speech (set of parts of speech depending on morphological features language: noun, adjective, numeral, verb).

The provided data sets allow not to store the set of all words of the language (for inflected languages it's pretty redundant) and form the words "on the fly" using basic word form and inflection. This significantly reduces the amount of information stored and allows us to perform determine the inflectional parameters of the input speech. Based on the data in provided sets, it is possible to generate the set of words and phrases of inflected language.

Dictionary of sign language contains a notation of gestures, their emotional state and subjects in which gestures are most often used. This allows to form a set of gestures.

The relation between the grammatical structures of inflected and sign language sentences is submitted as a generalized grammatical structures these sentences.

On one hand, the generalized grammar structures of inflected languages include:

1. a number that reflects the position of the element (word) in sentence;
2. inflectional parameters of words that can be used in this position;
3. parts of speech, which may or may not be used in this position;

4. On the other hand—generalized grammar structures of sign language include;
5. a number that reflects the position of the element (sign) in sentence;
6. the corresponding item number in inflected language sentence;
7. parts of speech corresponding to gestures of SL sentence;
8. Statistical database is used to analyze sentences already translated. It considers;
9. subject of the sentence;
10. grammatical structures;
11. frequency of sentences translation.

Every translated sentence is stored in a statistical database and has assigned a unique (ID) number. After the translation, this number is associated with a number of SL sentence for high speed translation of often translated sentences. In case of the translation prediction, user is given several options for translation. User-selected options are saved to improve translation quality in future. The sentences, with poor translation results are also stored and marked for further processing by authorized persons. The technology of the system-aided translation is given in Fig. 3.3:

The information technology of aided translation consists of the following steps:

1. The input sentence is searched for phrases. The phrase is a set consisting of a sequence of words.
2. The inflectional parameters are defined for each of the words and phrases of the input sentence.
3. Generalized grammatical structure is determined based on the parameters of inflection (part of speech and inflectional category) for each word of the input sentence. Generalized structure of inflected language contains a number that reflects the position of the element sentences, inflectional parameters of words and parts of speech that may or may not be used in this position.
4. If the generalized structure is not found, the search is performed for similar structures.
5. After finding generalized structure of translation that meets the input sentence, we get the corresponding generalized natural sign structure.
6. The search of "word → gesture" combinations are made for each word of input sentences.

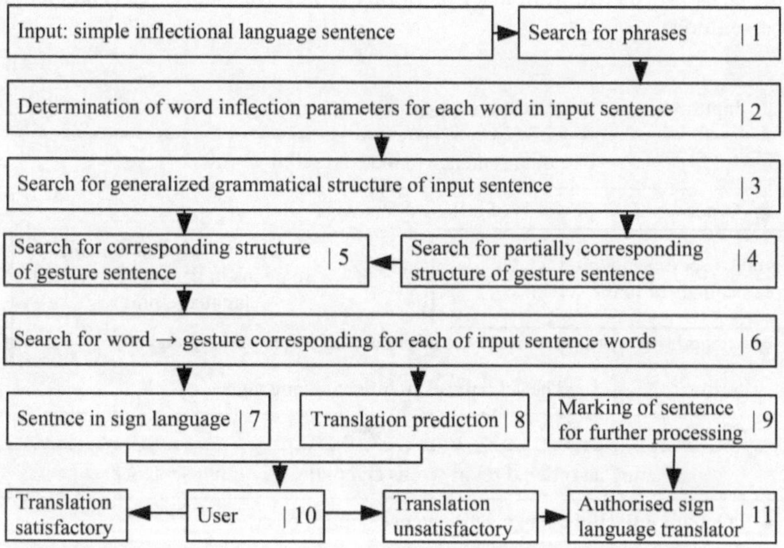

Figure 3.3. The block diagram of the of automated translation system technology.

7. If generalized structure of inflected language sentence is found, the corresponding generalized SL structure is selected, all compliances "word → gesture" are found, then the resulting sign language sentence is formed and output result is presented.
8. If generalized sentence structure of inflected language which overlaps with the structure of the input sentence is found, an attempt to forecast the translation is made. The generalized sentence structure that partially coincides with the input is used. Next, SL sentence is formed, using matching "word → gesture".
9. If there are no generalized structures that overlap with the structure of the input sentence is found, the sentence is added to the statistical database for further processing. If for some words the input sentence appropriate gesture is not found, such a word is marked in statistical base for further processing.
10. The user chooses one of the options after receiving the result.
11. The result is satisfactory. The sentences are marked as correct and added to the statistical base.
12. The result was not satisfactory and ought to be corrected manually. Corrected sentences are stored in the statistical base and marked for revision and future decision-making.
13. In case of generalized structure sentences or "word → gesture" matching was not found or the result of translation is not satisfactory the authorized translator can add new translation structure, modify already existing structure, add "word → gesture" reference or add a new gesture.

The phrase is a syntactic unit formed by a combination of two or more words related to each other in content and grammatically. If any member of the sentence is a phrase than regardless of whether it is phraseological or free, the phrase has a certain semantic autonomy. When translating a sentence, we need to define phrases from the set of words and consider that when analyzing the grammatical structure of sentences. For phrases in a sentence search purposes the following workflow (see Fig. 3.4) is proposed:

1. sentence is divided into an ordered set of words using spaces and punctuation and served as an ordered set of words,
2. words that contain hyphens are selected from the set of words and their search set of phrases is made,
3. we select all pairs of consecutive words and search for phrases that matched pairs,
4. similar searches are made for phrases consisting of three or more words,
5. if a sequence of words found in a set of phrases, then this sequence is labeled as one element of a sentence.

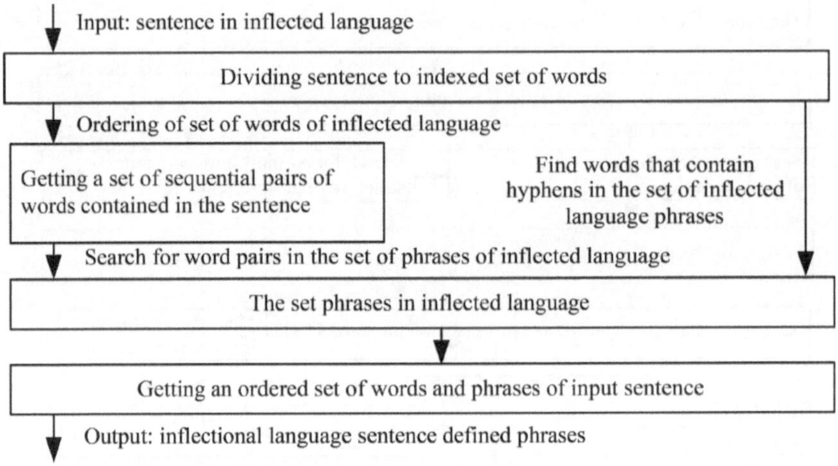

Figure 3.4. The block diagram, illustrating search for phrases in a sentence of inflectional language.

To determine the generalized grammatical structure of input sentence (according to the Fig. 3.5) the following workflow is applied:

1. querying statistical database if the sentence was translated before;
2. the inflectional parameters are determined for each word and phrase of input sentence using vocabulary of inflected language;
3. a search of generalized grammatical structures is made, for which the first element of a generalized structure includes inflection category corresponding to the first word of a sentence, the second element—the second word of each sentence;
4. if no appropriate grammatical structure is found the system searches for "similar" structures that overlap with the set: structures in which all parameters of all elements of inflection, except one match the input sentence.

The following functionalities are necessary to enable system scalability and improve translation quality by authorized persons:

1. statistics on the frequency of use of grammar structures for translation;
2. receiving and processing a set of sentences, for which translation result is unsatisfactory;
3. receiving and processing a set of sentences for which generalized grammar structures are not found, but translation is successful prediction based on similar structures;
4. viewing and editing correspondence between words of inflected language and gestures of SL.

To ensure processing of sentences that are translated incorrectly the next functional operators are required:

1. adding new generalized grammar constructions,
2. editing the existing structures,
3. editing the elements of generalized grammar constructions.

Functionality for editing generalized grammar constructions includes:

1. adding and removing sets of parts of speech and inflection parameters,
2. adding set of words that may or may not be used in the generalized grammatical constructions at certain positions.

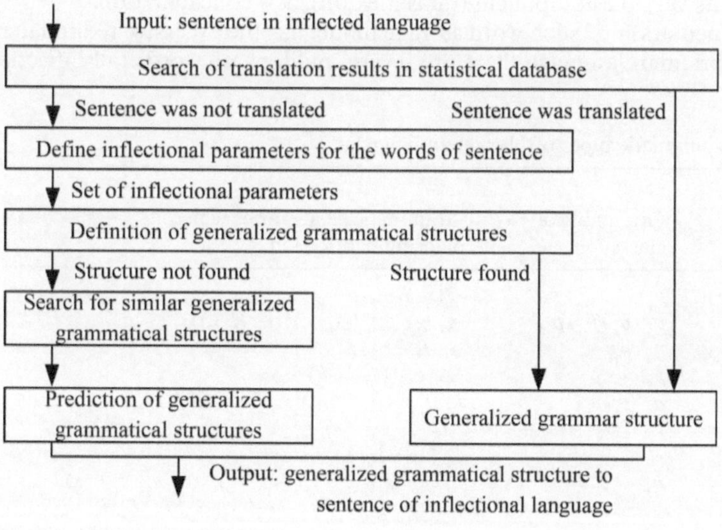

Figure 3.5. Block diagram with definition of the generalized grammatical structure for sentence in inflectional language.

To ensure the elaboration of sentences for which translation prediction was made the following functions are required:

1. creating new grammatical structures through design, used in translation prediction,
2. adding new generalized grammar constructions,
3. editing existing generalized grammar constructions,
4. editing elements of generalized grammar.

Presentation of dictionaries for translation

The grammar dictionary construction of inflected language is determined by the presence of inflection formal models. This means establishing and formalizing linguistic criteria according to which the entire set of words of the language is divided into specific subsets, whose mutual intersection of which is empty, and inside each inflection the same rules occurs. A subset of words with such properties are interchangeable paradigmatic types (Shirokov 2004).

By paradigmatic type we consider the group of tokens, the inflectional paradigm for which is characterized by the same number of grammatical forms, within which inflection is on the same (single) rule. For example, for Ukrainian language (which is the type of analytic-synthetic language), this means that the words that belong to a paradigmatic class have the same flexes in same inflections; corresponding to analytical forms built on the same model of formation (Shirokov 1998).

Categories of inflection or various languages may differ, depending on the language: for example, in the Ukrainian language there are four inflections, in Latvian – six, in Armenian – seven, while in most inflected languages only have one. Also, the most inflected languages don't have grammatical category of gender. These paradigmatic types exist for the Ukrainian language (see Table 3.1).

Based on the definition inflectional language words of inflected language are modeled as a combination of constant and variable components:

$$x = c(x)\,\&\,f(x), \tag{3.1}$$

where $c(x)$ – part of token x, which is in the process of inflection remains unchanged (quasi base), $f(x)$ – its variable component (quasi flexion), & – concatenation.

The implementation of such word storage model has the excessive redundancy. For example, in the Ukrainian language there are several millions of words (all inflection) and only

Table 3.1. Paradigmatic types of Ukrainian language.

Paradigmatic types	Grammatical classes	Grammatical categories which determine inflection	Number of grammatical meanings
T^1	p_1	$s*(i_1, K, i_7)$	14
T^2	$p_2 Y p_4 Y p_9 Y p_{10}$	$(g_1, g_2, g_3)*((s_1, s_2)*(i_1, K, i_7))$	24
T^3	$p_3 Y p_{11},$	(i_1, K, i_6)	6
T^4	p_5	$(g_1, g_2)*(i_1, K, i_6)$	12
T^5	$p_6 Y p_8$	$i_8, (i_{10}* s_1), (i_9* s_2), (i_{10}* s_2), (i_{12}* s_1),$ $((i_{12}, L, i_{17}) * s), i_{21}, i_{22}, i_{23}*(g_1, g_2, g_3)*s_1),$ $(i_{23}* s_2), i_{24}, i_{25}, i_{26}, i_{27}$	26
T^6	p_7	$i_8, (i_{10}* s_1), (i_9* s_2), (i_{10}* s_2), ((i_{12}, i_{13}, i_{14}) *s),$ $(i_{23}*(g_1, g_2, g_3)*s_1), (i_{23}* s_2), i_{24}, i_{25}, i_{26}, i_{27}$	18
T^0	$Y^{30}_{j=13}\, p_j$	i_{28}	1

about 100 thousand basic word forms. Instead of storing the entire word to database, we store an id of basic word form from the appropriate set of basic word forms, the number of positions in the word to which the base word form is unchanged and the id of flexion (the variable part of the word) from the appropriate set.

Given the characteristics of inflected languages (Shirokov 1998), (Shirokov 2004), (Shirokov et al. 2005) the dictionary of inflected language is shown as a set of parameters words:

$$W = \{W_i : W_i = \{I_{i_1} \in I, F_{i_2} \in F, k, In_{i_3} \in In\}\}, \tag{3.2}$$

where W_i – parameters of words of inflected language ($i = 0, \Lambda, N{-}1$; N – the number of words in the dictionary); F – the set of all possible word endings (inflections); k – item number in the basic word forms, where concatenation with flexion begins ($k = 0$ – for word forms completely different from the basic word forms); In – set of inflection parameters of inflectional language (for Ukrainian inflection parameters are number, gender, case, person, time, status, mode), I – set basic word forms inflected language:

$$I = \{I_i : I_i = \{word\,\mathrm{inf}, p \in P\}\}, \tag{3.3}$$

where P – a set of parts of speech; $word\,\mathrm{inf}$ – the base word form.

The phrases are considered as a sequences of words. The set of phrases can be represented as:

$$WS = \{WS_i : WS_i = \{word_j \in W\}\}, \tag{3.4}$$

where WS – set of all phrases, $i = 1, \ldots\ n$ – the number of words in the phrase, W – set the parameters of word.

The formation of sets W, F, I for all basic word forms of the inflected language shown in Fig. 3.6 using IDEF0 methodology.

IDEF0 – Function Modeling—methodology of functional modeling and graphic description of the process. Designed for formalization and representation of information flows within the system via interconnected functional blocks. IDEF0 peculiarity is its emphasis on hierarchical representation of objects that simplify understanding of the subject area.

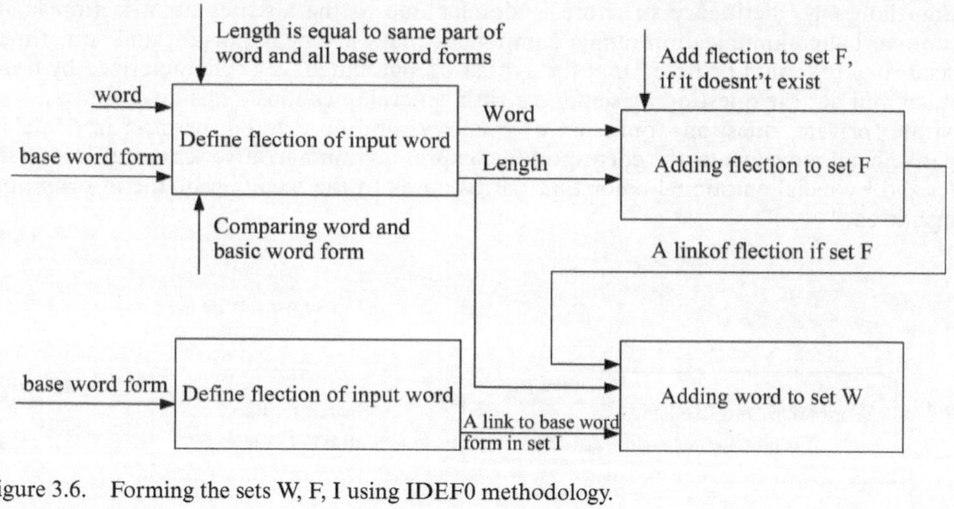

Figure 3.6. Forming the sets W, F, I using IDEF0 methodology.

In IDEF0 logical connections between information flows are considered, not the sequence of their execution time (IDEFØ Function Modeling Method).

Direct objective of this model is a simple task of getting full text of the speech on three parameters I_{i_1}, F_{i_2}, k, which is implemented, using respective operator (H):

$$H = \begin{cases} Left\left(I_{i_1}, k-1\right) \& F_{i_2}, & k \neq 0 \\ Left\left(I_{i_1}, 0\right) \& F_{i_2}, & k = 0 \end{cases}, \qquad (3.5)$$

where *Left*(*Word*, *Length*) – function for getting first *Length* symbols of word *Word*; & – concatenation.

Some difficulty arises for the inverse problem (H^{-1}): finding three parameters for existing words that define it clearly. Implementation of operator (3.5) for the database as a non-deterministic function leads to inability of indexing table. To solve this problem, the algorithm is proposed (see Fig. 3.7).

The input word is broken down into all possible combinations of constant and variable components, and the search is conducted on indexed fields of sets of basic word forms (I) and inflections (F):

$$H^{-1} = \left| \begin{array}{l} \text{i} = 0 \\ \text{l} = \text{length(Word)} \\ \text{continue while i} < \text{l} \\ \quad \text{Head} = \text{LEFT(Word, i)} \\ \quad \text{Tail} = \text{RIGHT(Word, li)} \\ \qquad \text{If exists base word form } \text{I}_{\text{i}_1} \in \text{I, that starts with Head} \\ \text{ and exists inflexion} \text{F}_{\text{i}_2} \in \text{F, that equals Tail, than} \\ \text{get values } (\text{I}_{\text{i}_1}, \text{F}_{\text{i}_2}, \text{k}) \\ \text{else} \\ \quad \text{i} = \text{i} + 1 \\ \text{end} \end{array} \right| \qquad (3.6)$$

Sign language dictionary structure is simpler due to the absence of inflection in it. In non-verbal communication mimic component plays a very important and sometimes crucial role. It should be noted that the syntax of sign language is characterized by non-manual marker: the questioning sentences with general questions—raising eyebrows; in a separate (private) question—omission of eyebrows and head tilted forward; in denial—negative head movements, the corresponding expression; in narrative sentences it is characterized by facial emotional color that corresponds to the meaning of the information transmitted.

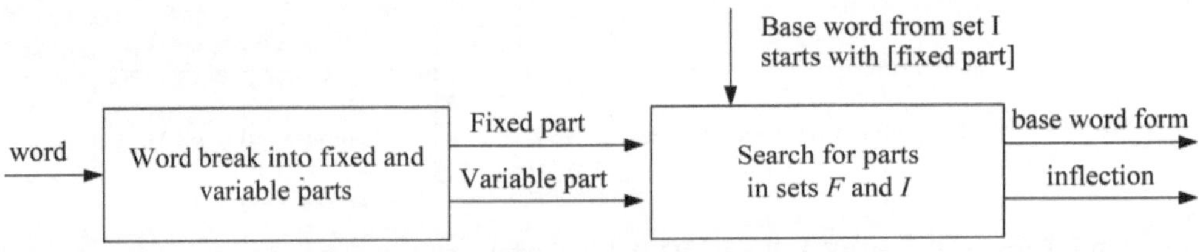

Figure 3.7. Block diagram with search of word parameters from sets W, F, I.

The set of gestures is modeled as:

$$Ges = \{Ges_i : Ges_i = \{word\ des, pges \in PGes, em \in Em\}\}, \tag{3.7}$$

where *word des* – description of gesture, $PGes \in \{p_1, K, p_n\}$ – corresponding part of speech, $Em \in \{em_1, em_2, em_3\}$ – a set of emotional coloring; em_1. – narrative emotional coloring, em_2. – questioning emotional coloring, em_3 – other emotional colorings.

Presentation of translation grammar structures correspondences

When interpreter translates from one language to another mutually we receive unambiguous pairs of structures that convey meaning: sentence in input language → appropriate sentence in the output language. Suppose that these structures pairs can be represented as a certain generalization. After analyzing certain amount of sentences pairs obtained by translation, we can get a generalization by fixing the order of words in a sentence. The generalization instead of specific words in the sentence has the sets of words that can be used at these fixed locations. Thus, in relation to the total sentences, the small amount of grammar structures for translation is received.

The sequence of words of the language and punctuation is understood as the sentence. The sequence of words of the language related to the parts of speech that convey meaning (function words of sentences is rejected) is considered as grammatical sentence structure. The grammatical sentence structures, based on the number of words in them are also distinguished. Elements of grammar sentence structures (unlike sentences) are the sets of words in sentences that will be used at fixed positions. This approach is presented in the Fig. 3.8:

$$\begin{aligned} Gr &= \{Gr_i = \{word_j \in W \| seq_j \in Seq \| p \in P, In_{i_1} \in In, num\}, \\ GStr_i &= \{ges_j \in Ges \| gseq \in Gseq \| pges \in PGes, num, gesnum\}\}, \end{aligned} \tag{3.8}$$

where *num* – number of element in structure of inflectional language, *gesnum* – number of element in structure of sign language, *In*- set of inflectional parameters of a language, *P* – a set of parts of speech of inflectional language, *PGes* – a set of parts of speech of sign language, *W* – a set words of inflectional language, *Ges* – a set words of gestures.

$$Seq = \{Seq_i : Seq_i = \{word_j \in W : word_j \in W, n\}\}, \tag{3.9}$$

$$Gseq = \{Gseq_i : Gseq_i = \{ges_j \in Ges : ges_j \in Ges, n\}\}, \tag{3.10}$$

where *n* – number of word in set of words.

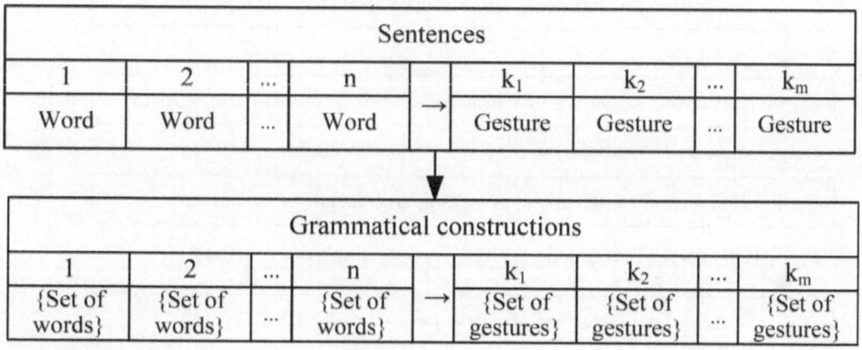

Figure 3.8. Grammar constructions of translation.

Formation of the set of grammar structures based on bilingual corpus of sentences of inflectional language and sign language is as follows (Fig. 3.9):

1. in the set of sentences contained in the corpus we search for sentences that overlap. If distinction between words is only one of the elements, only one sentence ought to be chosen;
2. conduct grouping of sentences and get a sequence of words that can be used at fixed positions in each of these sentences;
3. conduct analysis of sequences, determine whether each sequence of words can be used in every sentence, find exceptions, add to sequence other similar words, which can be used in this position;
4. after analyzing the sequences of words a grammatical sentence structure is formed;
5. add grammar constructions to sentences set, again search for sentences and grammatical structures that overlap to expand grammatical structures. For example, having a corps of sentences:
    a. Він має брата. (He has a brother)
    b. Вона має брата. (She has a brother)
    c. Вона любить брата. (She loves a brother)
    d. Вона любить кота. (She loves a cat)
6. conduct their generalization. The first sentence overlaps with the second and third sentence, but is different from the fourth. As the result of such synthesis, instead of getting the first three sentences, the grammatical sentence structure is achieved:
    a. {Він, вона…} {Має, любить…} брата.
7. This grammatical structure overlaps with the fourth sentence of the corps, allowing to expand it to {він, вона…} {має, любить…} {брата, кота}. Repeat this step, for as long as grammar will not cease to expand;
8. search for grammar structures in which sequences overlap and provide analysis of sequences for grammar structures expansion.

After receiving a set of grammatical constructions we analyze each of the sets of elements that go into it. By generalized grammatical constructions we understand a construction that

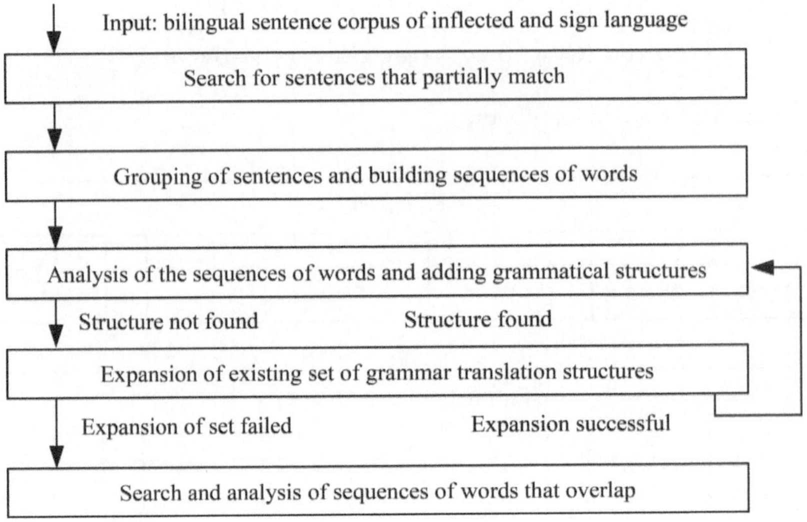

Figure 3.9. Block diagram with forming the set of grammar structures.

contains a combination of inflectional parameters of a language instead of set words, given by expression (3.11):

$$GGr = \{GGr_i = \{p_j \in P, In_j \in In, num\}, \\ GStr_i = \{pges \in PGes, num, gesnum\}\}. \quad (3.11)$$

Generalized grammatical structures contain the sets inflectional parameters of inflected language on fixed positions. For SL in generalized grammatical structures we also store the position of the element in the generalized grammatical structures of inflected language (see Fig. 3.10).

Elements of generalized grammatical structures may be union or intersection of sets of certain inflectional parameters. Generalized grammatical structure can contain elements of certain part of speech, or exclude them. Additionally, to the sets of inflectional parameters generalized grammatical structure can contain a sequence of words which can or cannot be used. Examples of elements of generalized grammar structures are given for the Ukrainian language in Table 3.2:

Formation of the set of generalized grammar structures (Fig. 3.11) based on grammar translation is as follows:

1. for each grammar sentences structure a generalized grammatical structure is built: for each set of words inflectional parameters of the language common to all of its elements are defined;
2. after forming generalized grammatical structure, check its accuracy based on sentences from bilingual corps;
3. after forming a set of generalized grammatical structure analysis is performed to define the structures that overlap (based on parts of speech and inflectional parameters that are used) and its possible merge;

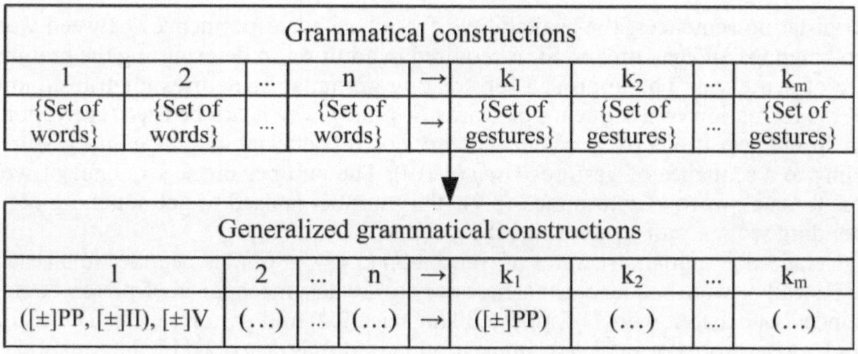

Figure 3.10. Generalized grammar structures of translation.

Table 3.2. Examples of elements of generalized grammatical structures.

Element generalized grammatical structure	Description of the element generalized grammatical structure
$((p_1),(s_1),(g_1),(i_1))$	Nouns singular masculine nominative
$((p_1, p_2),(s_1),(g_1),(i_1))$	Numerals and nouns singular masculine nominative
$((p_1),(s_1),(g_1),())$	Nouns singular masculine any case
$((p_1),(s_1),(g_1),(-i_2,-i_3))$	Nouns singular masculine any case other than generic and dative
$((p_1),(s_1),(g_1),(i_1)$, + (він, вона))	Nouns singular masculine nominative and the set of words {він, вона}
$((p_1),(s_1),(g_1),(i_1)$, – (один, два))	Nouns singular masculine nominative except the set of words {один, два}

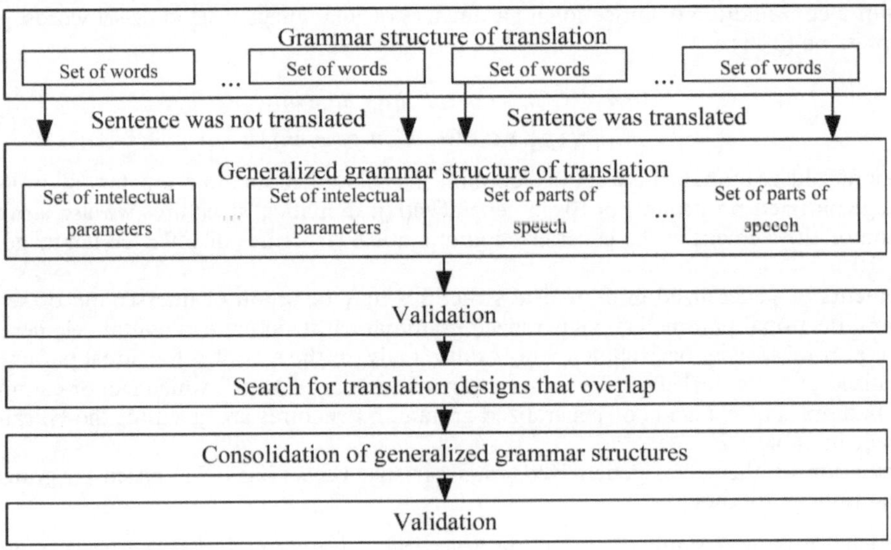

Figure 3.11. Formation of set of generalized grammar structures of translation.

4. if we get a new generalized grammatical structure due to merging, check its accuracy based on sentences from bilingual corpus that match the design.

Obtaining correspondences between words of inflected language and gestures of sign language

When translating sentences, the availability of a set of correspondences between words of inflected language and gestures of SL is required in addition to determining the grammatical structure of sentences. The approach for creating grammar structures for translation (3.8) involves the formation of a sequence of words *Seq* (3.9), which can be used (correspondence between specific positions of words in sentences of inflectional language and gestures corresponding to a sequence of gestures *Gseq* (3.10)). The number in the sequence of words in sentences for each word is determined (3.9). This number is used to get matching gesture in corresponding sequence of gestures for the given word.

For generalized grammar structures of translation (3.11), grammatical inflectional categories are used instead of word sequences. It is necessary to create a mechanism of presentation of the correspondences between words of inflected language (3.2) and gestures (3.7). The set of relationships between words of inflected language and gestures is described by the equation (3.12):

$$Corr=\{Corr_i : Corr_i = \{word_j \in W,\ ges_k \in Ges\}\} \tag{3.12}$$

To create a set of relations, chosen sequences of words and gestures are used:

1. for each grammatical structure the list of sequences and gesture sequences is defined;
2. for each words in the sequence of words, the pairs "word → gesture" are searched;
3. if the set of matches doesn't contain a pair of "word → gesture", add it to the set.

The process of forming a set of correspondences between words of inflected language and gestures is shown in Fig. 3.12.

If the formation of the set of correspondences between words and gestures (3.12) gives several translations for a single word, it is necessary to analyze the grammar structure (3.8), containing the sequence with these words. For generalized grammar structures (3.11), obtained

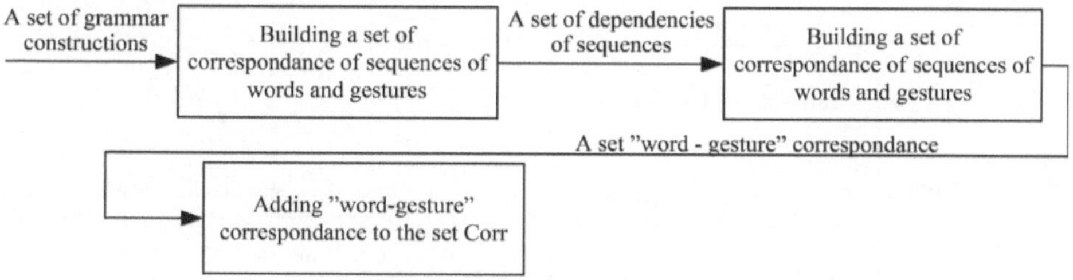

Figure 3.12. Building a set of correspondence between words and gestures.

from grammar structures (3.8), containing words with variability of translation, it is necessary to add or remove a set of inflectional categories or sequences, to select the correct translation.

For example, the word of Ukrainian language «**ходити**» (walk) can be translated into sign language differently, depending on the context: the result of the translation of sentence «**людина ходить по вулиці**» (Man walking on street) and «**місяць ходить по небу**» (Moon goes through the sky). Similarly, to display the word «**ходити**» different gestures are used. To ensure the right choice of gesture, a new generalized grammatical structure of translation ought to be added. The above sentence corresponds to generalized grammatical structure of translation:

$$((p_1,p_{11}),0,0,(i_1)),\ ((p_6),0,0,(i_9,i_{10},i_{11})),\ ((p_1,p_{11}),0,0,(i_6)) \rightarrow \\ \rightarrow ((p_1,p_{11}),1),((p_1,p_{11}),3),(p_6,2),$$

this design It is processed as follows, to ensure proper selection of gesture:

$$((p_1, p_{11}),0,0,(i_1)-(\textit{місяць})),\ ((p_6),0,0,\ (i_9,\ i_{10},\ i_{11})+ \\ -(\textit{ходить,ходив...})),\ ((p_1, p_{11}),0,0,(i_6))\rightarrow\ , \\ \rightarrow((p_1, p_{11}),1),((p_1, p_{11}),\ 3),(p_6,\ 2) \tag{3.13}$$

and then:

$$(\textit{місяць}),\ (\textit{ходить,ходив,...}),\ ((p_1, p_{11}),0,0,(i_6))\rightarrow\ . \\ (\textit{МІСЯЦЬ},1),\ (\textit{ХОДИТИ ПО НЕБУ(МІСЯЦЬ)},\ 3),\ (p_6,\ 2) \tag{3.14}$$

As the result, sentence «**людина ходить по вулиці**» get generalized grammar structure (3.13) and sentence «**місяць ходить по небу**» – (3.14), providing the right choice of gesture for word «**ходити**».

Generation of the set of generalized grammar structures of translation

The availability of relationships mechanisms representation between grammatical structures of inflected and sign languages and the relations between words of inflected language and gestures allows to perform automated translation from inflected to sign language. The necessary condition for design and application of automated translation system is bilingual corpus of sentences in inflectional and sign language. Completeness of the corpus affects the quality of proposed automated translation system.

To ensure sufficient completeness of bilingual corpus we use the program used in schools for children with hearing impairments. This program includes gestures and sentence them evenly cover all the topics of daily (household) of signed communication. The program accumulates the knowledge of different subjects to study sign language at the same level at which native verbal language is studied.

The crucial feature of the program clearly reveals its main tasks: mastery of the mother/first language, forming the base for mastering a second language (verbal) and the study of two cultures: the deaf and hearing communities.

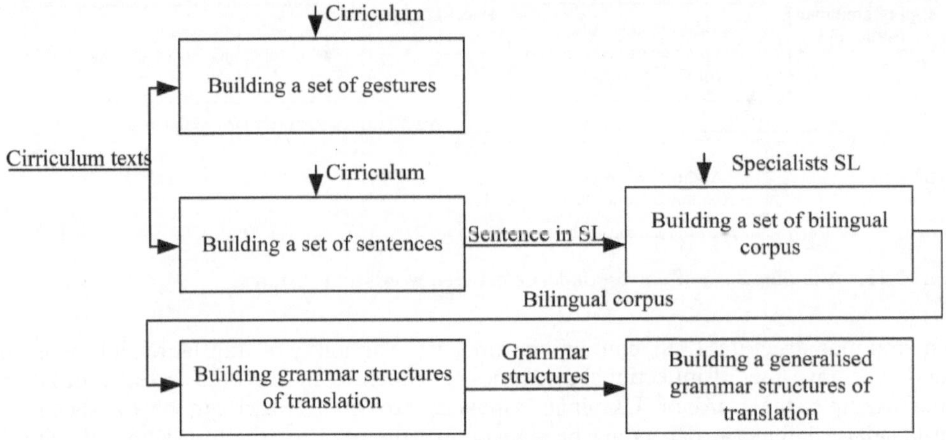

Figure 3.13. Building a set of generalized grammar structures of translation.

As a result of the development program, sign language students:

1. obtain the sign vocabulary of various subjects,
2. fill sign vocabulary by learning new concepts,
3. can reasonably express their views,
4. are able to build grammatically correct sentences.

The automated translation system using the corpus that provides a fairly complete grammar and gestures reserve, necessary for everyday communication.

For generating the set of generalized grammatical structures the following technology, presented in the Fig. 3.13 is proposed:

1. a set of gestures is formed using program for learning sign language that covers common topics of everyday communication,
2. based on texts from the program the set of sentences in inflectional language are formed, which are used in the study of sign language,
3. on the basis of software, using signed language specialists' assistance, bilingual corpus is formed: each sentence inflected language is associated with a sentence of sign language, words of inflectional language sentence are in matched with gestures in SL sentences and the order of words in inflected sentences and gestures gesture sentence is fixed,
4. based on the bilingual corpus a set of grammar sentences structures (3.8) containing a sequence of words and gestures is formed,
5. as a result of analysis of grammar sentences structures (3.8) a set of generalized grammar structures is formed (3.11) and correspondences between words and gestures are built (3.12).

3.2 ALGORITHMIC MODELS OF AUTOMATED INFORMATION TECHNOLOGY OF TRANSLATION FROM INFLECTIONAL LANGUAGE TO SIGN LANGUAGE

To ensure feasibility of the proposed in the previous section approaches (models and charts for conversion of information), due to translation from inflected to sign language, the algorithmic models are required for submission of:

1. dictionaries of inflected and sign languages,
2. correspondence between the grammatical structures of inflected and sign languages,
3. correspondence between gestures and words (as infological models),

4. algorithmic functionality for building information technology,
5. algorithmic functionality for use of information technology,
6. algorithmic functionality for extensibility of translation.

For algorithmic implementation dictionaries of inflected and sign languages, Relational Database Management System (RDMS) was chosen for correspondences between the grammatical structures of sign and inflected language, correspondences between words and gestures as infological model. Dictionaries of inflected and sign languages are presented as tables and relations between them. This allow to form words of inflected language as combination of basic word forms and inflections, as well as to determine inflectional parameters and part of speech of input word, to define parts of speech and emotional coloring by gesture designation.

Proposed infological models should apply algorithmic functionality for operation and subsequent extensibility of information technology for translation of inflected to sign language.

The database structure for building of informational technology

To implement the proposed mechanisms of submission of dictionaries of inflected and sign languages, correspondence between words and gestures, correlations between grammatical structures of inflected and sign languages using RDMS was proposed. The dictionary of sign and inflected languages is shown as tables, containing sets of words, gestures and their options. The relations between the grammatical structures of sign and inflected languages are shown as tables, containing generalized grammatical structures of sentences inflected language, generalized grammatical structures of sentences inflected language and relationships between data tables.

Infological model of dictionaries of inflected and SL is presented as follows (Fig. 3.14).

The table "Dictionary of inflected language" contains a set of words in inflected language (W), sets of inflections (F) and basic word forms (I) – tables "Flexions" and "Basic word forms" respectively. The reference to grammatical categories which determine inflection is stored for each of the words: depending on the morphological features of speech data of grammatical categories can be number, gender, case, person, time, status, mode and other. The set of gestures is found in the table "Gesture dictionary". The table contains reference to the word, which currently affects gesture and theme, in which the gesture is often used. Sets of basic word forms contain a reference to table "Parts of speech" to determine the words belonging to certain part of speech. To apply grammar structures (3.11) the following infological model is proposed (Fig. 3.15).

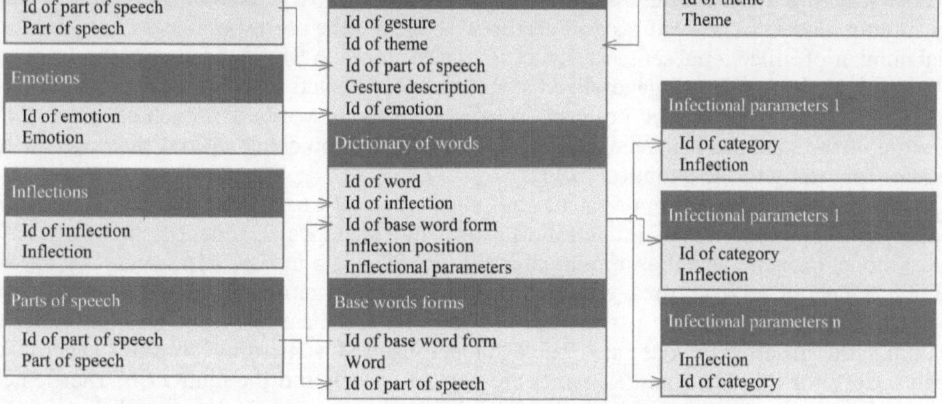

Figure 3.14. Infological model of dictionaries of inflected and sign languages.

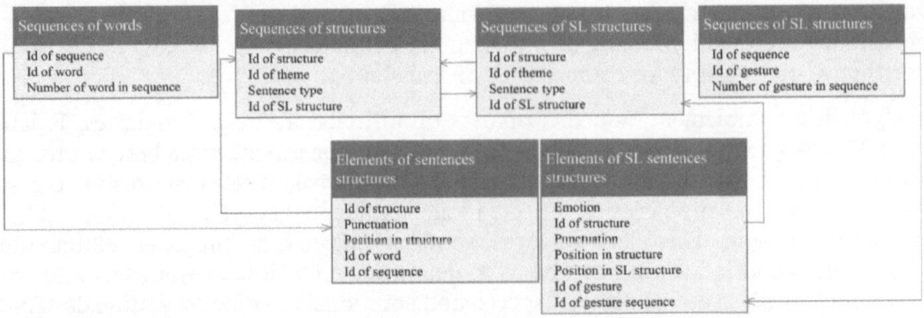

Figure. 3.15. Infological model of sentences structures of inflectional and sign languages.

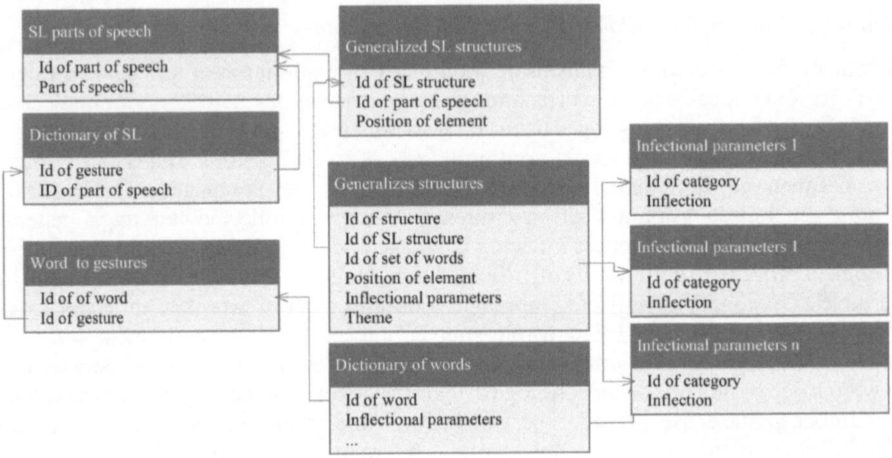

Figure 3.16. Infological model of generalized grammar structures of inflected and sign languages.

The structures of sentences are stored in the table "Sentences structure", the structure of sentences of SL—in "SL structures". These tables contain information about the type of topic sentences and sentences. "Elements of sentence structures" can contain both certain words (id of word) and a sequence of words (id of sequence). To preserve the order of elements of sentence structure each item has a serial number. "Elements of structures of SL sentences" may also contain sequences of gestures and gestures. To ensure the correspondences in translation serial number of inflectional sentence structure is also stored in SL sentence structure.

Through analysis, a set of generalized structures of inflected and sign languages containing only category of inflection, not certain words or sets of words. For each element of the structure of the sentence has been allocated a set of inflection category and the order of the element (according to the equation (3.11)).

Creating a generalized grammar structures allowed to get the structure of the SL sentence by analyzing the categories of inflection of words in a sentence and identificator (Id). Infological model for generalized grammatical structures is shown in Fig. 3.16.

Table "Generalized structures" contains the generalized structures of sentences of inflected language. The main elements of this table is the inflectional category and serial number of the item in the structure of sentences. Table "Generalized SL structures" contains generalized constructions for SL. The main elements are part of speech and the number of the element in the structure. To use this infological model following algorithmic functionalities are proposed: obtaining the id of generalized structure for sentence in inflected language, obtaining the id of generalized SL structure using the id of generalized structure.

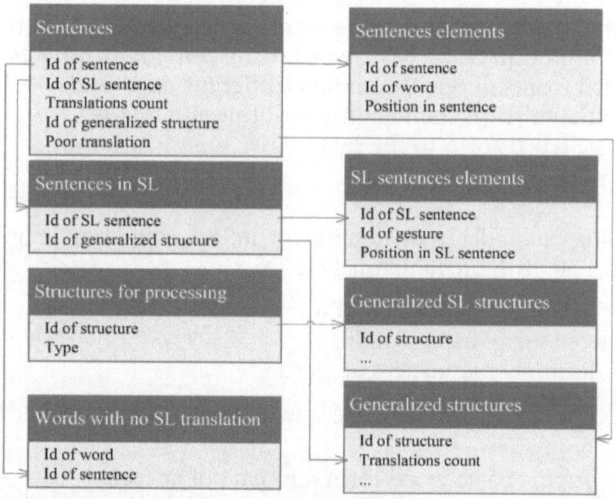

Figure 3.17. The block diagram of statistical database structure.

Statistical database, presented in the Fig. 3.17. contains tables for translation, that allows to:

1. ensure faster processing sentences translated before,
2. study words for which sign matches were not found,
3. handle sentence, for which generalized grammatical structure of translation was not found,
4. study grammar translation, received automatically as prediction result.

Table "Sentences" contains sentences in inflectional language. The id of appropriate SL sentence, the id of used generalized grammar structure and translations count is stored into table if the sentence was translated before. If the translated sentence obtained poor results, it is indicated for the processing in the field "Poor translation". The table "Generalized structures for processing" contains id's of translate designs that require processing, depending on the type divided into grammatical structures:

1. derived from translation prediction,
2. that have no proper corresponding sentence construction in SL,
3. with poor translation result obtained.

Table "Words without appropriate gesture" contains id of sentences in which they are used, a number of words from the table "Dictionary of inflected language" that were used in the translation of sentences, and were not found in the table "Compliance of gestures to words".

Functional models for building informational technology

To provide the desired functionality within the information technology it is necessary to offer the following algorithms:

1. construction of vocabulary of sign and inflected languages,
2. construction grammar structures of translation from inflected to sign language,
3. obtaining correspondences between words inflected gestures and sign language.

Models for building informational technology

To form the set W (words of inflectional language), the set of F (inflections) and the set of I (base word forms) we propose to implement the next approach (Fig. 3.6) using algorithm (3.15):

Input: basic word form and all its word forms (grammatical meanings) Length (common part for basic word forms and all its inflections) For all word forms define the ending (different part) and add it to the set F (if such ending is not in set already) Add to basic word form to the sets I Form set W	(3.15)

To build the infological model of dictionaries of inflected and sign languages there is need to offer the following algorithmic functionality:

1. adding a basic word form, inflection and word,
2. obtaining basic word form, its inflection,
3. determine the inflectional parameters of word.

Adding basic word forms to the set of basic word forms (3.3) is as follows (3.16):

Input: basic word form and part of speech Determine the [id of part of speech] from table "Parts of speech" Add [basic word form] and [id of part of speech] to table "Basic word forms"	(3.16)

The process of adding a base form, from to the set of word forms using IDEF0 methodology is shown in Fig. 3.18.

Adding words and inflections to the corresponding sets is the done using following algorithmic scheme (3.17):

Input : word of inflected language, basic word form,
parameters of inflection
Define [id of inflectional parameters 1],
[id of inflectional parameters 2], ...
[id of inflectional parameters n],
[id of basic word form] by selection
from tables "Inflectional parameters 1",
"Inflectional pa - rameters 2", "Inflectional parameters n",
"Basic word forms"
Define k = position of word, starting from
which it differs from basic word form
l = word length
flexion = RIGHT (word, l)
If flexion not empty
        If exists [flexion] in table "Inflexions"
 define [id of flexion]
        Else add flexion to table "Inflexions" and get its id
        End if
Else define number of empty flexion from table "Inflexions"
End if
Add [id of basic word from], [id of flexion],
[position of flexion], [ids of inflectional categories],
to table "Inflectional language dicionary"
	(3.17)

The process of adding words and inflections using IDEF0 methodology is shown in Fig. 3.19. The inflection position are defined when adding word its inflectional parameters,

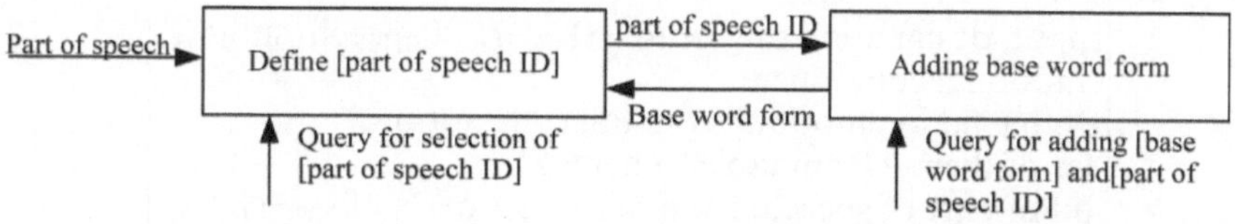

Figure 3.18. The block diagram of adding base word form to database.

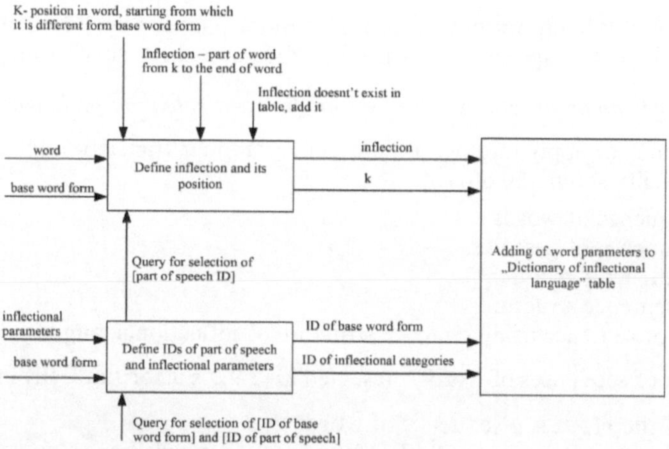

Figure 3.19. Process of adding word and inflection to database.

corresponding to basic word form. The id is determined in case the inflection already exists in the database, otherwise—it is added to the table.

Getting basic word forms, inflections, inflectional category is given by (3.18):

Input: word in inflected language
Define [id of base word form] for word from
table "Inflectional language dictionary" and [base word form]
from table "Base word forms" using linked field
[id of base word form]
Define [id of inflexion] for word from
table "Inflectional language dictionary" and [inflexion]
from table "Inflexions" using linked field [id of flexion]
Define [id of inflectional category n] for word from
table "Inflectional language dictionary" and
[inflectional category] from "Inflectional categories n"
using linked field [id of inflectional category n] (3.18)

To define parts of speech of input word, the next algorithmic sequence is suggested by (3.19):

Input : word in inflected language
Define [id of base word form] for word
from table "Inflectional language dictionary"
For defined [id of base word form] get [id of part of speech]
from table "Base word forms" and [part of speech] from
table "Parts of speech" by linked field [id of part of speech] (3.19)

To add a database gestures and their options, the following algorithm is proposed below:

```
Input: designation of gesture, SL part of speech, subject,
emotional component
Define [id of emotion] from table "Emotions",
 [id of theme] from table "Themes",
 [id of part of speech] from table "SL parts of speech"
Add [gesture definition], [id of emotion], [id of theme]
and [id of part of speech] to table "Gestures dictionary"
```
(3.20)

Algorithms of obtaining options of gesture require selection of [id of emotion], [id of subject] and [id of part of speech] from the table "Gestures dictionary" using their id.

Construction of grammar structures of translation from inflected to sign language

Copy the template. To apply infological models of grammar translation the following algorithmic functionality should be offered:

1. creating a sequence of words,
2. creating a sequence of gestures,
3. creating a sentence structure,
4. creating SL sentence structure,
5. obtaining id of sentence using sentence structure of inflectional language.

Creating a set of sequences of words, presented in (3.9), is algorithmically implemented as:

```
On the input is a sequence of words
Create a record in the table "Sequences of words",
which contains the first word of sequence and its serial number
Get [sequence id]
For each of the following words adding [sequence id], [word],
 [number of word in sequence] to table "Sequences of words"
```
(3.21)

Creating a set sequence of gestures (3.10) is analogous operation the previous one. To create the grammatical structure of the sentence next algorithmic sequence is given by (3.22):

```
The input to the algorithm is the sentence structure
consisting of words or
sequences of words, type and subject of the sentence
Create a record in the table "Structure of sentences"
that contains
[type of sentence] and [topic] get [id of sentence structure]
for add-ed record
For each item sentence structure
If the item is the word
    define [word number] from the table "Dictionary of
    inflected language"
    Add the record to the table "Elements of sentence structures",
    which includes [id of sentence structure], [id of word],
    [punctuation] [position of word in the structure]
End if
If the element is sequence of words
    search for it in the table "Sequences of words"
    If the sequence is found define [id of sequence]
    Else create a new sequence using algorithm described above
    and determine [id of sequence]
Add the record to the table "Elements of sentence structures",
which includes [id of sentence structure], [id of sequence],
 [punctuation] [position of sequence in the structure]
End if
```
(3.22)

Creating a structure of SL sentences is similar to the creation of the structure of sentences of inflected language, except that in the table "Elements of the structure of signed clauses" contains [id of structure gesture sentence] and [id of structure of the sentence] to ensure consistency between the elements of sentences of inflected and sign languages. After creating a structure of SL sentences, it is stored in corresponding sentence structure of inflected language, in "Structure of sentences" table.

On the basis of the curriculum for children with hearing impairments, the set of sentences in inflectional language was formed. The sentence is submitted as a sequences of words. Translation of this set of sentences is performed with experts on SL speech assistance. Each word of the sentence is linked to corresponding gesture. As a result of the translation of each sentence of the training set bilingual corpus of sentences was obtained.

For getting the id of inflectional language sentence the following algorithm (3.23) is proposed:

```
The input to the algorithm is the sentence in
inflectional language
For each word of sentence find id in the table
"Dictionary of inflect-ed language"
Form the sequence of indexed numbers [sentence]
Find the list sentence structures ids [list], for which in the
table "Elements of the structure of sentences" this
[sentence][1] is a sequence of words or in the field [word id]      (3.23)
and [position in the struc-ture] = 1
For all next words of sentences structures [list] look for those
that contain [sentence][i] in the sequence of
words or in the field [word id]
and for which [position in the structure] = i
If this structure is found - return its number
```

Based on bilingual corpus the set of grammar constructions of translation was formed, using the algorithm (3.24):

```
For each sentence [sentence1] from bilingual corpus
Search for grammar structure [structure],
where all elements except one are the same as in
structure of [sentence1]
If structure [structure] is found extend the elements
 which are different
Else
Search for [sentence2], in which all elements except one      (3.24)
are the same as in [sentence1]
If [sentence2] is found create new grammar construction
 [structure] based on [sentence1] and [sentence2],
which contains parts of sentences
that are the same, and sequence of words that are different
for [sentence1] and [sentence2]
End if
End if
```

After receiving the set grammar structures with help of a specialist in signed language the analysis and corrections were conducted. Incorrect grammar constructions were labeled and stored in a separate set. Regarding to irregular translation structures, the constructions algorithm (3.24) is repeated, until the set stop increasing.

To create a generalized grammar structure of translation (3.26) the following algorithmic functionality (3.25) is offered:

For each [struct] from table "Sentences structures" For each sequence of words in [struct] define inflectional parameters which are the same for all elements Search for grammar sentence structure in table "Generalized grammar structures", which contains defined inflectional parameters If such structure doesn't exist, add new structure [GenStruct] to table "Generalized grammar structures" End if For [struct] search for corresponding [gesturestruct] from table "Structures of SL sentences" Define parameters of gestures which are the same for all elements of [gesturestruct] and define generalised SL structure If defined [gesturestruct] already exists, add its id to [GenStruct] Else add [gesturestruct] to table "Generalized SL structures" and a reference to its id to [GenStruct] End if	(3.25)

Getting correspondences between words of inflected language and gestures of sign language

In addition to determining the sentence structure it is necessary for translation to have correspondence between words and matching gestures. Using the correspondence between the structures of sentences and structures of SL the set of correspondences "word → gesture" was formed. Table "Word to gesture" contains the ids of words and ids of corresponding gestures.

To get the matching "word → gesture" for each of the sentences in the table "Structures of sentences" and matching SL structure from table "Structures of SL sentences" correspondence between word sequences and sequences of gestures were obtained. Using data according to the table "Sequences of words" and "Sequences of gestures" was formed table "Word to gesture". In case of non-unique correspondence of "word → gesture" this infological model does not allow to define a clear relationship between the set of words and a set of gestures. If one word corresponds to several gestures for selecting the correct gesture the relationship between the sentences' structures and SL structures is used. If the few words are corresponding to one gesture difficulties in obtaining compliance "word → gesture" does not arise. In addition, for such gestures we can get a list of synonyms. To fill the table "Word to gesture" the following algorithm was used (3.26).

For each sentence structure from table "Sentences structures" define the corresponding structure of the "Structures of SL sentences" Each sequence of words in the sentence structure is matched to corresponding sequence of gestures For each word determine its sequence id and gesture id from corresponding sequence Add id of word and id of gesture to the table "Word to gesture"	(3.26)

Functional model of informational technology usage

The following algorithmic functionality should be offered, to work with dictionaries infological model:

- defining basic word forms of word,
- defining inflection for word,
- defining part of speech,
- defining parameters of gesture,
- defining inflectional parameters of word.

Algorithms to implement this functionality (3.18)–(3.20) were developed during the construction of dictionaries.

To work with infological model of generalized grammar structures of translation we should offer the following algorithmic functionality:

- defining the id of generalized grammatical structure for sentence in inflectional language;
- editing of existing generalized grammatical structures;
- obtaining generalized grammatical structure of translation by its id;
- identification of generalized grammatical SL structure of sentence using id of generalized sentence construction of inflectional language.

The following algorithmic functionality (3.27) is proposed to determine the id of generalized grammatical structure of input sentence:

```
For each word of sentence determine its inflection parameters
Choose from the table "Generalized structure" generalized grammatical
 constructions of sentences in which the number of elements is equal
 to the words count of a sentence (the maximum [number of elements]
for structure = number of words in a sentence) and generalized
grammatical construction elements include parameters
corresponding to inflections of word in sentence
```
(3.27)

Determining the id of generalized SL grammatical structure, is done by selecting [number of generalized sign design] from the table "Generalized structures", using the id of generalized construction of inflectional language.

To search for phrases in a sentence the algorithm (3.28) is proposed:

```
Input: sentences in inflected language
Create element of class CSentence sentence based on input
sentence
Using method sentence. LoadSentence() the input sentence
is transformed
 to indexed sequence of class CWord words
CWord list[] = null;
k = count of words sentence.words, i = 0, j = 0
Repeat until i < k-1
   If sentence. words[i]. fullword + '' + sentence. words [i+1]. fullword
exists in table "Phrases"
   list[j] = CWord(sentence. words[i]. fullword +'' +
 sentence. words[i + 1]. fullword)
          i = i + 2
          j = j + 1
          Else i = i + 1
End cycle
Resulting list contains indexed sequence of words
and phrases: sentences. words = list
```
(3.28)

To form the gesture sentences based on sentence of inflectional language proposed the following algorithm (3.29):

For each element of sentence Sentence. words define corresponding gesture Sentence. words. GetGestureId() and generate a set of gestures ids Create element of class CGestureSentence Gsentence Use Sentence. Structure. ElementsCorrespond to define the order of gestures in SL sentence based on gesture positions and form Gsentence. Gestures	(3.29)

The algorithm of information technology for automated translation consists of the following steps (compare to Fig. 3.3):

1. Input sentences are verified for existence of phrases.
2. Using dictionary of Ukrainian language, parameters inflection, part of speech, basic word form, inflection, and number in the database table are defined for each of words and phrases from the table "Dictionary of Ukrainian language".
3. Based on inflection parameters the generalized grammatical construction is determined (3.30):

Input: CSentence sentence Using method GetSentenceStructure(), in the table "Generalized struc-ture" search for generalized sentence structure that corresponds to the input parameters: the number of elements = number of elements of the sentence, the first structure element contains inflection category, the relevant to categories of inflection of sentence. words [1], the second category includes element of inflection of sentence. words [2], for each word of sentence	(3.30)

4. If generalized grammatical construction if not found the search of partially corresponding constructions is performed (3.31):

Input: CSentence sentence Using method GetPartialMatch(), in table "Generalised structures" search for structure that partially matches sentence on the input pa-rameters: i = 0 Repeat, until i < sentence. words. count Search for generalized structure for which elements count = sentence elements count, all elements of structure except sentence. words [i] correspond to the input sentence	(3.31)

5. If generalized grammatical construction of input sentence is found, corresponding generalized grammatical construction of SL sentence is determined.
6. Using "word → gesture" define gestures corresponding to input words.
7. If we find a generalized sentence structure of inflected language, the corresponding generalized SL structure and all compliance "word → gesture", then the resulting sign language translation is displayed.
8. In case of partially corresponding grammatical construction an attempt to predict translation is made. Generalized structure of sentence which overlaps with the input partially is used. Based on that structure using "word → gesture" the SL sentence is built. The result-

ing structure is added to the database as inactive and labeled in the table "Generalized structures to handle".

9. If no partially corresponding grammatical construction is found the sentence is marked for further processing by authorized sign language expert. The sentence is added to the table "Sentences" without the specified number of generalized structure. If for some of the words of a sentence appropriate gesture is not found, the word is denoted to statistical base table "Words without appropriate gesture" for further processing.

After receiving the translation user chooses one of the following:

1. The translation is satisfactory. The sentences are marked as correct and is added to the statistical base. The id of sentence, id of SL sentence and construction id are stored in the statistical database table "Sentences".
2. The translation is unsatisfactory. The id of sentence, id of structure used for translation (if received) are stored in statistical database table "Sentence" and are set to "Poor translation" for further processing.
3. If the structure is not found, matching "word → gesture" not found, or the result of translation is not satisfactory—authorized translator can add a new translation design or change already existing, a match for "word → gesture" or a new gesture.
4. The information about the system is formed based on the statistical database.

Authorized persons are able to:

1. get statistics on the frequency of use of generalized grammar structures,
2. view a list of sentences that were often translated in the system,
3. view a list of sentences which indicated poor quality of translation;
4. view a list of sentences for which during translation no generalized grammatical structure of translation was found;
5. view a list of generalized transfer structures requiring updates;
6. add new generalized structures of translation or edit existing ones.

The required number of structures is selected from the tables "Frequency of translation" and "Generalized structures" on a bound field [id of generalized structure] sorting by the field [number of uses] for getting the frequency of usage of grammar structures.

Information about the sentences that were translated in the system is stored in the database table "Sentences". Field [Translations count] has the amount of translations for sentence. [id of generalized construction] is set on first translation and can be used, if the sentence is re-translated to ensure greater speed of system. [Unsatisfactory translation] is set if user submitted result as poor translation. If the sentence that was translated has no [id of generalized structure] set, then this sentence needs a check. Authorized persons are able to obtain lists of sentences that require identification, design or adjustment of generalized translation correction.

3.3 EXPERIMENTAL TESTING OF INFORMATIONAL TECHNOLOGY OF TRANSLATION FROM UKRAINIAN TO SIGN LANGUAGE

Implementation of the proposed information technology for automated translation of inflected languages to sign language was considered on example of Ukrainian. To test proposed in the previous sections approaches, algorithms, models and charts, to obtain the translation system from Ukrainian to sign language experimental applications were implemented for:

1. construction of Ukrainian and sign language dictionaries,
2. forming the set of correspondences between Ukrainian and grammar of sign language,
3. testing of the effectiveness of translation.

Building the dictionaries of Ukrainian and sign languages

To achieve the translation Ukrainian and SL, according to infological model (Fig. 3.14) dictionaries were developed. The basis of Ukrainian dictionary was electronic dictionary "Словники України on-line" (Dictionaries of Ukraine on-line). To form the vocabulary of gestures of Ukrainian sign language, the complex program "Ukrainian sign language" was used to master sign language (Program Complex). The database was filled using developed software application.

The input of application had the set of words in Ukrainian with the right inflection parameters obtained from the dictionary "Dictionaries of Ukraine on-line" (Dictionaries of Ukraine on-line). Each word was applied in pair with its basic word form. Using operator (3.7) the input word was divided into a combination of quasi base and quasi flexion and using algorithms 3.15–3.17 were added to the database.

Based on the educational program of the "Ukrainian Sign Language" used in schools for the people with impaired hearing to master sign language sets of SL sentences and gestures were formed. Each gesture had its designation assigned. Gestures were grouped according to the topics of the curriculum in which they are used. The given set of gestures and their options have been added to the database.

As a result of the work on filling database the following numbers were obtained:

1. base word forms – 140440,
2. Ukrainian words in all word forms – 1971342,
3. structural word endings of words – 2067,
4. gestures of SL – 3155.

Bilingual corpus of sentences building

Bilingual sentence corpus is necessary to form the set of correspondences between both grammar Ukrainian and sign language. To ensure the efficiency of obtained grammar structures the sentences that are used in different situations, with different grammatical structure and covering most of the vocabulary and grammar structures for everyday communication ought to be contained.

The set of sentences in Ukrainian were formed that are used in the study of SL using SL study program for specialized schools. All of this sentences were decomposed into simple sentences. The sentences were grouped according to the topics of the curriculum.

The set of correspondences between words and gestures was formed based on the bilingual sentences corpus of Ukrainian and sign languages. Using the developed software based on the algorithm 3.25 the set of words and gestures that correspond to them was formed. For each word, which is found in sentences of bilingual corpus, the set of all word forms was allocated and gesture compliance was set.

Bilingual corpus contains 9800simple sentences with appropriate vocabulary, having a appropriate stock, which covered the situation on the possible speech topics studied in the program of study Ukrainian sign language specialized schools. Number of correspondences between Ukrainian words and gestures was 35298 records. Number of correspondences between the basic word form and gestures equals 2754 records

Corpus for generation of generalized grammar structures of translation

After receiving bilingual corpus of sentences, obtained correspondences between sentences in Ukrainian and SL ought to be summarized. For this purpose, inflectional parameters (number, gender, part of language and a generalized category that includes case, identity, time, status, mode) were proposed. The first stage was the formation of grammar structures, based on sets of words. In the second phase, achieved sets were summarized by the relevant parameters of inflection as well as generalized grammar structures of translation were formed.

Then, the set of grammar structures of translation was formed, using the developed software (based on algorithms 3.22–3.24). Partly grammar structures were formed by combining the sentences into grammar structures.

There were received 2101 grammar translations, from which 293 derived generalized grammar structures of translation. This allowed the automated translation of sentences from corpus.

The automated translation efficiency verification

One of the crucial thing was to test the effectiveness of the developed solution. To elaborate this the set of simple sentences in Ukrainian was used. There was condition, that these samples ought to be used in everyday communication between people with hearing impairments on one hand, but they should not belong to the previous study of the proposed solution—on the other hand.

The set of simple sentences in Ukrainian was formed, derived from electronic publishing for the deaf. To test the effectiveness of the translation software was developed. On output the software formed a set of SL sentences grouped by the following criteria:

1. successful automatic translation,
2. automated translation based on generalized grammar structure partially matching the input sentence,
3. sentences for which translation could not be predicted,
4. sentences for which not all words were translated to gestures.

The set of sentences in Ukrainian and SL derived from the translation was submitted for validation by specialists of SL.

When conducting test translation on set of Ukrainian sentences taken from bilingual educational corpus 100% automatic translation was obtained. There were 500 simple sentences selected from chosen publication to conduct practical test of the technology. As a result of translation 321 (64%) sentences were translated automatically, while 119 (24%) were not translated due to the lack of word to gesture matches. Precisely 60 (12%) were not translated because of the lack of generalized grammatical structures (for 33 (55%) of these sentences generalized grammatical structures that overlap with the input sentence were proposed). The test results are presented in the Fig. 3.20:

As a result of the expansion of vocabulary and establishing correspondences between words and gestures we received results of 423 (85%) sentences translated automatically and 31 sentences (6%) with automated translation. The total percentage of successful translation was 91%. The test results shown in Fig. 3.21:

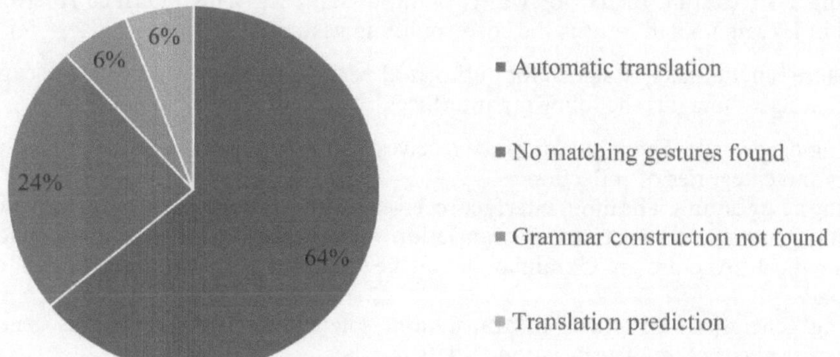

Figure 3.20. The results of the translation on sentences from periodicals tests.

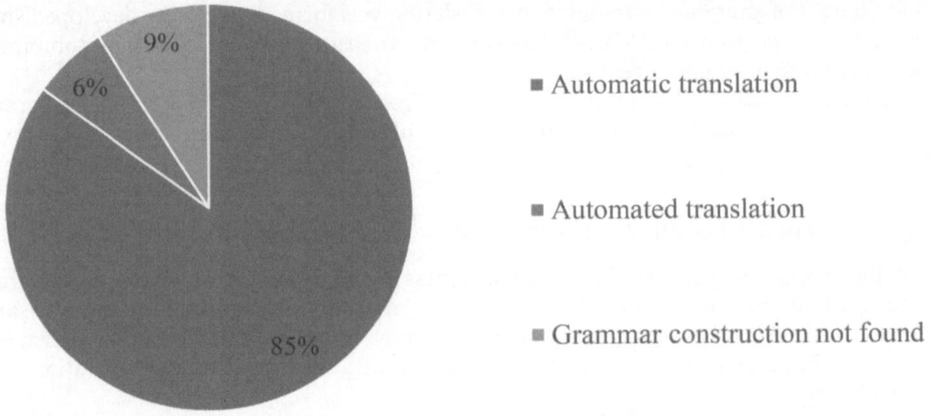

Figure 3.21. Testing of the translation on sentences from periodicals on vocabulary extension.

Web-service implementation

Further development of this solution implies its use on different platforms as a service. Designed Web-service API provide access to the following features of automated translation system:

1. function to perform translation,
2. features for editing and expanding the existing system,
3. function to get statistical information about the system,
4. function of system administrator.

Functions of translation give the ability to use the translation algorithm or only receive information about the grammatical structure of sentences and correspondence between words and gestures to allow using own translation algorithms. The main interface maintains:

1. Obtaining SL sentence by the input sentence in Ukrainian. This interface reads a simple sentence in Ukrainian and using described above (see Fig. 3.3) translation algorithm returns SL sentence.
2. Receiving inflectional parameters of words for input sentence. This interface receives input word in Ukrainian and returns its inflectional parameters.
3. Getting generalized grammatical structure corresponding to the sentence. This interface reads a sentence in Ukrainian, returns generalized grammar structures of Ukrainian and sign language sentences.
4. Getting word-gesture match for words of input sentence. This interface receives input word in Ukrainian and returns the corresponding gesture.

To ensure functionality, designed for authorized persons engaged in editing and expansion of the existing system has the following interfaces.

1. Adding new words. The input interface receives Ukrainian language word, its basic word forms and categories of inflection.
2. Adding new gestures. The input interface receives gesture description and its characteristics.
3. Adding generalized structures of translation. The input interface receives generalized grammatical structure in Ukrainian language and the corresponding design of sign language.
4. Editing generalized structures of translation. The input interface receives generalized grammatical structure of translation and its number.

5. Creating a word-gesture match. The input interface receives the Ukrainian word and the corresponding gesture.
6. Display of a list of generalized transfer structures derived from translation prediction. The interface returns a list of generalized grammar and translation and their ids.
7. Displays a list of generalized designs of translation data over time. The interface returns a list of generalized grammatical structures according to the set time interval.
8. Displays a list of sentences for which a "poor translation" is selected. The interface returns a list of sentences, grammatical constructions that were used for these sentences and their ids.

Functions for statistical information on the database include the following interfaces:

1. Display of information about the usage of grammatical constructions. The input interface receives the number grammatical structure and parameters that indicate the type of information required. On the output, depending on the specified parameters, we obtain the number of uses of grammatical structure, a list of proposals for which was used this design, the number of uses of this design in predicting the transfer, the number of poor translations using this design.
2. Display of a list of grammatical structures that often or less often used. The input is the number of designs to display and sorting options. The output interface gets a list of structures, the count of translations and their ids.
3. Display of a list of sentences that often or less often used.
4. Display of the usage of translation system for a certain period of time.

For the realization of the administrator of the system developed following interfaces:

1. Adding or removing users.
2. Setting or removing permissions for groups of users.
3. Viewing recent events for particular user.

3.4 CONCLUSION

This section discusses the inflected language to sign language translation solution. It constitutes an important element for the new systems for the training and communication with deaf people.

The model of inflected and sign languages dictionaries representation was improved. It significantly reduces the amount of memory required to store dictionaries. Unlike the existing approaches, the words of the language are stored in the database as a combination of appropriate indexed sets of basic word forms and inflections, by which the word is automatically generated.

The elaborated mechanism for representation of relations between words of inflected language and gestures of sign language is quite innovative. It allows to align the Ukrainian words with corresponding gestures. The advantages of this representation are both the ease of creating, editing words—gestures relationships between and the possibility to form these relationships using text corpus.

For the first time the mechanism of generalized grammar translation structures for inflected and sign languages was developed, allowing to define the grammatical structure of the input sentence and an appropriate structure of gesture sentence for translation.

In reference to popular existing approaches, presented solution uses inflectional parameters of the words of input sentence to determine its grammatical structure.

The developed ability to continuously expand and improve the quality of translation stands innovative approach in such automated translation area. The advantages of this technology include: relative simplicity (for a translation system based on rules) of its expansion and enhancement of quality of translation as a result of self-learning.

The information technology for nonverbal communication of deaf people was further developed by inclusion of automated translation functionality, allowing to empower its practical applications. The advantages of this expansion include ability to create interactive inflected language translators to sign gestures with speech playback using three-dimensional actors as realistic animation.

Experimental software was implemented to build and Ukrainian and sign language dictionaries, to build a set of correspondences between Ukrainian and sign language grammar, to test the effectiveness of translation. The last stage considered web service that implemented the functionality of automated translation based on the proposed information technology.

REFERENCES

Barmak, O.V. & Krak, Iu.V. & Romanyshyn, S.O. 2014. Automated translation from inflectional languages to sign languages. International Journal Information Content and Processing. Vol. 1(1): 37–43.

Barmak, O.V. & Romanyshyn, S.O. 2012. Computer Technology for Translation from Ukrainian to Sign Language. Bulletin of Ternopil National Technical University. Nr 1(65): 195–208.

Barmak, O.V. & Romanyshyn, S.O. 2013. Implementation of Information Technology for Automated Translation from Flective Languages to Sign Language. Bulletin of Vinnitsa Polytechnic Institute. Nr 5: 134–142.

Dictionaries of Ukraine on-line [Electronic resource]. ©ULIF, 2001–2014. http://lcorp.ulif.org.ua /dictua/.

Dugast, L., & Senellart, J. & Koehn, P. 2007. Statistical post-editing on SYSTRAN's rule-based translation system. Proceedings of the Second Workshop on Statistical Machine Translation. Association for Computational Linguistics: 220–223.

IDEFØ Function Modeling Method [Electronic resource]. Knowledge Based Systems, Inc, 2010. http://www.idef.com/idef0.htm.

Krak, Iu.V. & Barmak, O.V. & Romanyshyn, S.O. 2011. Generalized Grammar Constructions for Automated Translation from Ukrainian to Sign Language. Artificial intelligence. Nr 3: 136–146.

Krak, Iu.V. & Barmak, O.V. & Romanyshyn, S.O. 2013. Implementation of Information Technology for Translation from Flective Languages to Sign Language. Speech technologies. Nr 1: 21–29.

Krak, Iu.V. & Barmak, O.V. & Romanyshyn, S.O. 2014. The Method of Generalized Grammar Structures for Text to Gestures Computer-Aided Translation. Cybernetics and Systems Analysis. Vol.50(1): 116–123.

Krak, Iu.V. & Romanyshyn, S.O. 2014. Automated translation from Ukrainian to sign language: algorithmic implementation. Programming problems. Nr 2-3: 285–294.

Program Complex "Ukrainian sign language" [Electronic resource]. 2005. http://www.mon.gov.ua/education/average/programs_gluh.

Senellart, J. 2006. Boosting linguistic rule-based MT system with corpus-based approaches. GALE PI Meeting. Boston: 33–55.

Shirokov, V.A. & Bugakov, O.V. & Griazuhina, T.O. 2005. Corpus linguistics: Monograph. Ukrainian Lingua-Information Fund of NAS of Ukraine. Dovira: 472–483.

Shirokov, V.A. 1998. Information theory of lexicographical systems. Dovira: 331–342.

Shirokov, V.A. 2004. Phenomenology of lexicographical systems. Naukova dumka: 327–339.

CHAPTER 4

Interoperability of the health care information resources

Y.G. Kryvonos
V.M. Glushkov Institute of Cybernetics of NAS of Ukraine, Kyiv, Ukraine

O. Khorozov
Institute of telecommunication and global information space of NAS of Ukraine, Kyiv, Ukraina

A. Smolarz & W. Surtel
Lublin University of Technology, Lublin, Poland

A. Kalizhanova
al-Farabi Kazakh National University, Almaty, Kazakhstan

INTRODUCTION

The Health Information and Management System Society define interoperability in healthcare as: "(…) the ability of different information technology systems and software applications to communicate, exchange data, and use the information that has been exchanged. Data exchange schema and standards should permit data to be shared across clinicians, lab, hospital, pharmacy, and patient (…)" (HIMSS 2013). Information exchange is a key factor towards the faithful execution of medical workflows. However, the diversity of clinical data turns the information exchange process into a challenging issue of integration. This lack can lead to inefficient medical treatment and can raise healthcare costs. Interoperability is achieved in part by the use of consistent standards that define the syntactic and semantic meaning of information. Implementation of standards can reduce the costs and timelines of health technology development projects. There exist a variety of standards for healthcare information exchange today. Health Level 7 (HL7) standards (http://www.hl7.org) are a popular choice for both local communities and regional eHealth initiatives. The mission of HL7 is to provide standards for the exchange, management and integration of data that support clinical patient care and the management, delivery and evaluation of healthcare services. There is still no unique technical standard covering all aspects of health data communication. Nevertheless, HL7 International organisation has released a series of standards which are the most widely used by the current healthcare systems, and has defined guidelines that target on integration of healthcare components, from medical devices to management systems.

Health Information System (HIS) is in broader sense any system used for collecting, storing and retrieving data on activities and phenomena relating to the health care sector. This system can be based on the Information Technology and Communications (ICT) infrastructure. Generally the goal of an HIS is to contribute to better efficiency and quality of provision of health care by providing reliable and relevant information on treatment of patients, for management of the health care sector at all levels and efficient allocation of financial and human resources.

An HIS is mostly designed for Electronic Health Records (EHR) and information resource formation. EHR include Basic Patient Information; patient administration (Admission Note; Transfer; Discharge Summary; Episode Note); Pharmacy order & Prescription; Medical Record; Imaging Report; Lab Test; Referral Letter & Referral Summary; Payment Sheet. Application Programming Interfaces (APIs) of an EHR management system should support the standard protocol for information exchange between laboratory, radiological, telemetry, pharmaceutical and health care insurance systems for semantic interoperability of healthcare domains (Krivonos and Khorozov 2014).

The transport protocol is using eXtensible Markup Language (XML) representation of the message structures and domain-specific vocabularies (e.g. HL7). The problem of compatibility of medical, administrative and statistical information is related to integration of different types of data in accordance with an HIS model and the structure of Clinical Document Architecture (CDA). Integration of healthcare applications cannot be achieved without standardising the exchange of information. In this context, the focus is on communication standard HL7 XML for data exchange.

The requirements for standardisation are focused on the functionality of the software that use a common data model, structured messages and the capability of importing XML-files to the database (DB) of a medical reports management system. Relational technology of XML data storage is associated with two types of files, "data-centric" and "document-oriented". In the former case it applies parsing and relational presentation data for subsequent analysis. In the latter case, XML-files are loaded into the database as character entities (Khorozov 2014) or a hybrid EHR database model is used for the storage of XML-sections of the complex CDA together with the relational representation of the document context. In this case we are dealing with API implementation and EHR database formation.

Fast Health Interoperability Resources (FHIR) is an interoperability standard developed for the exchange, integration, sharing and retrieval of electronic health information to provide a framework and related standards known as HL7 International. Today in HISs, the common standard is Consolidated CDA, which is designed to transfer entire documents, rather than a simple list. This means that when a physician requests information about a patient, the system needs to transfer multiple documents to fulfil the request. FHIR, on the other hand, makes it simple to receive a specific piece of requested information. This approach is much more efficient for systems exchanging specific information, rather than entire documents. A major technology change embodied in FHIR is a move away from a document-centric approach to a data-level access approach using well-defined APIs.

The article deals with APIs of heterogeneous applications in healthcare based on an FHIR server for management information resources, parsing HL7 XML-format messages and patient Electronic Medical Record (EMR) formation. A database model is proposed which uses a combination of XML-storage sections of EMR together with a relational representation of the document context. For the class of XML-files, which contain structured parts and sections with complex elements, data access is achieved through the use of XML Schema, XML indexing and XPath query. We consider the actual application for a patient's vital signs telemonitoring system framework and Patient Medical Records (PMR) formation.

The main objective is to achieve interoperability between e-Health organisations for easier access to healthcare information. The proposed approach addresses integration of a wide range of heterogeneous data sources and a decision-making support process through OLAP (OnLine Analytical Processing) and data-mining facilities. It builds a suitable solution for integration of high volumes of healthcare data. This approach is a step towards decentralisation and strengthening of interoperability in the healthcare sector, which results in a better service for the patient and medical staff. The goals aimed for are interoperability and data access.

4.1 STRATEGY FOR HIS DEVELOPMENT

The HIS for the health sector should not be built as one monolithic information system. The result would be definitely very vulnerable to any changes in the sector and, with poorly defined interfaces, hampering future development of the system. We rather advocate for a system with clearly defined distinct components co-operating and exchanging data through defined and well documented data interfaces that are developed. General requirements for designing information flows (medical, administrative, financial and statistical) in the healthcare industry are related to system functional class.

The requirements apply to the following key functional classes:

1. Reference Information Resources for information support of medical facilities and management.
2. HIS intended for information support of processes of diagnosis, treatment, rehabilitation and prevention of patients.
3. Statistical Information System (SIS) for health management.

A long-term HIS strategy is necessary to achieve objectives such as:

1. developing and managing HIS components inside the health sector;
2. a set of standards for development of HIS components;
3. a set of standards for data exchange between HIS components (formats and protocols) and maintaining the definitions of these standards;
4. a common or distributed approach to technological platforms;
5. responsibility for collection of specific data items and basic data flows between distinct institutions to prevent duplications and ensure completeness and consistency of collected data;
6. a procedure for maintenance of a global HIS strategy.

The ICT infrastructure and the corresponding application software can optimise expenditures only through adherence to the explicitly formulated HIS strategy in the health care sector.

The development of the HIS should adhere as much as possible to the standards. These standards could be classified as technological standards, communication standards or coding/classification standards.

The **technological standards** specify hardware, software and database platforms for implementation of an HIS. These standards will be broadly accepted international formal or "de facto" standards. However, a choice of available options may be selected for financial reasons to harness advantages of open licenses that can be negotiated with major vendors for the health care sector.

The **communication standards** for the health care domain should definitely be used for data exchange between different institutions acting in the health care sector. These standards can be advantageously utilised as a foundation for internal communication within an institution. The following standards of this class can be recommended: a) HL7 XML protocol; b) CEN standards for electronic prescriptions; c) ISO PKI technical specification; d) ISO standard on health cards.

The **classification standards** enable unique identification of different phenomena, qualities and objects within the health care domain. The following coding/classification standards may be used: a) International Classification of Diseases (ICD-10); b) Coding Atlas for Pharmaceutical Substances (CAPS); c) Diagnostic Related Groups (DRG); d) Medical Service Classification.

4.1.1 *Interoperability standards in healthcare*

The HIS at the level of a health care facility generally provides the following functionalities: registration of patients; management of contacts and referrals; reporting; claiming and invoicing; financial accounting and book-keeping. These functionalities are provided by ITC support as well as standards and APIs in order to achieve integration of applications, and to inter-operate with other healthcare providers. In order to effectively respond to the interoperability needs, a care unit should set out an Integration Roadmap, showing how it will use open APIs and standards to migrate towards the Integrated Digital Health Care Record.

Interoperability of the HIS is classified into three categories: Syntactic; Semantic; Device. Syntactic interoperability is provided by using standard data models (RIM) for communication among healthcare systems. Syntax is a grammar to convey semantic structure. Semantic interoperability is provided by using terminology standards such as SNOMED, LOINC, ICD-10 for understanding clinical terms. Device interoperability is provided by using specifications for medical devices and links observations with terminology. Integration standards from HL7 can form a key part:

1. HL7 Version 2 is a standard use most appropriate for message based interactions (ordering, prescribing and database synchronisation).
2. HL7 Version 3 is a standard most appropriate with sophisticated structures for complex interactions and EHRs, and better suited to national level with data interactions.
3. CDA is the electronic format for healthcare and planned clinical documents.
4. FHIR is an open standard API for medical applications. FHIR offers an easy way to get systems connected using off the shelf web standards. Right now, it would make a great basis for a newly constructed interface for mobile systems. FHIR is designed to encompass the capabilities of all of the other standards.

A fully integrated patient record system across all the HIS is in order, where the meaning 'integrated' is that vital patient information and clinical decisions can be viewed by an authorised user in any single instance. The objective is integration of information from different ITC systems. The HL7 produces an electronic system to messaging standards as well as such standards as electronic document structure and content standards to support systems interoperability (Clinical Data Standards in Health Care: Five Case Studies, 2005). Those categories of standards are presented in Appendix A.

4.1.2 *Taxonomy of HIS components*

The components of the HIS can be classified in the following way:

Clinical component is used for support medical personnel's operations and duties. It is usually built as a patient centric component to store and retrieve data on patients and the history of their treatments. It provides a health record for each individual that shows every health-related event. The health record should show all significant events for the patient (made the diagnosis, performed the procedure, determined the outcome of a treatment). The subcomponent of the clinical component can be PACS (Picture archving and communication system) that enables recording and retrieval of image data on patients or telemonitoring the vital signs of a patient.

Surveillance (monitoring) component collects observations about the health status of a population. Surveillance component is used to track epidemics and reportable diseases like TB or HIV/AIDS. Such components are often used with socio-economic and demographic statistics to make projections and comparisons.

Managerial component is used to manage resources for a facility, district or region. It presents demographic statistics, service delivery statistics, and financial records and produces reports showing information such as utilisation, revenue, expenditure, efficiency and accessibility. Clinical, surveillance and financial components provide input data for the managerial component.

Financial component gathers data necessary for claiming and accounting for the rendered health care. The clinical component may provide data for the financial component.

4.2 THE FHIR SPECIFICATION CONCEPT

FHIR is considered to be an emerging HL7 standard that takes advantage of the positive elements of its ancestors HL7 v2 and HL7 v3. The basic building block of FHIR is "Resource", meaning a basic set of structured data. A resource is an entity, instance of data, which is either stored or exchanged. For example, a resource could be defined as a medication list, a problem list or lab results. In today's system, standard coding sets such as LOINC for lab results, or RxNorm for medications, allow software applications to exchange just the data that is needed and present it to clinicians or consumers.

In FHIR specification the data are not sufficiently "tied down" to semantic consistency. To achieve consistency in syntax and semantics a set of FHIR adoption profiles is needed. Such profiles have general incorporating Meaningful Use vocabularies such as SNOMED CT, RxNorm, and LOINC, resource definition specific profiles for concepts like "quantitative lab results," "vital signs" and "medication allergies".

4.2.1 *Contents of the resource*

Every resource has an identified version and a known identity, a URL which can be used to address the resource. Resources contain a human readable Hypertext Mark-up Language (HTML) representation of the content, a set of data structured and represented in XML or JSON format. Resources have the following optional or mandatory elements and properties: a base set of defined data elements specific to the Extensions type; a human-readable narrative description of the contents of the resource; metadata—information about the resource; tags—labels affixed to the resources which can be used to define additional operational behaviour such as workflow, security, etc.

4.2.2 *Implementing FHIR resource profiles*

Problems statement

1. FHIR profiles are generalised—multiple implementation options will not be interoperable.
2. Interoperability relies on criteria, e.g. singular invariant XPath per concept.
3. Some resources are undefined as to how they are constructed in detail (e.g. Blood Pressure).
4. Can existing metadata standards such as XML be used to define the profile to the right detail?
5. Can existing mature tooling be used to implement FHIR validation?
6. Can we integrate the terminology?

Interoperability Criteria

1. Payload defined and controlled by computational metadata (not documents).
2. A concept must have a singular invariant XPath—defined at different levels.
3. Values must be of the same dimension (units).
4. Codes for a concept must be from the same Value Set.
5. Detailed XML schemas need to be shared between interoperable exchange partners.

Remark: The Blood Pressure component instances (systolic and diastolic) can be expressed in different ways according to the XML schema.

4.2.3 *Conformance in FHIR*

The conformance in an FHIR specifies the resource type interactions that are supported by the FHIR Server for a specific resource. The FHIR API defines the resources as a set of operations also known as interactions on the resources, where individual resources are managed by their type. The FHIR server must always define conformance for the resources by specifying the list of possible operations.

The following are the list of operations at various levels of the interaction:

Instance Level Interactions

Read – specifies the current state of the resource;
Update – updates an existing resource by its ID;
Delete – deletes the resource;
History – retrieves the update history of a particular resource.

Type Level Interactions
Create – creates a new resource with a server assigned ID;
Search – searches the resource type based on some filter criteria;
Validate – checks whether the content would be acceptable as an update;
History – retrieves the update history for a particular resource type.
Whole System Interactions
Conformance gets a conformance statement for the system;
Transaction – updates, creates or deletes a set of resources;
Search – searches across all resource types based on some filter criteria;
History – retrieves the update history for all resources.
In addition to the above-listed basic operations, there are Document endpoints to specify conformance for a specific resource.

4.2.4 *Resources for medical documents*

FHIR resources can be used to build documents that represent a composition: a set of coherent information that is a statement of healthcare information, particularly including clinical observations and services. A document is a set of resources with a fixed presentation that is attested by humans, location and devices. Documents may be exchanged between systems and also persist in DB storage and management systems. Applications claiming conformance to this framework claim to be conformant to "FHIR documents". FHIR documents may be PMR (focused on patient healthcare information). HL7 will develop profiles in the future giving additional guidance on appropriate representation of clinical documents (Consolidated CDA). Note that FHIR defines both document format and a reference resource. An example of resources for CDA mapping are represented in Table 4.1.

Table 4.1. List of the resources for CDA mapping.

Resource	CDA Mapping
General:	
CarePlan	Act, Organiser, Clinical Document
ReferralRequest	Act, Encounter, Procedure, Clinical Document
Procedure	Procedure, Observation, Act
Medications:	
Medication	Manufactured Product (Substance Administration), Product (Supply)
MedicationPrescription	Act, Supply, Substance Administration
Diagnostics:	
Observation	Observation, Organiser
DeviceObservationReport	Observation, Organiser, Clinical Document
Attribution:	
Patient	Patient Role, Related Subject
Practitioner	Assigned Author, Assigned Entity, Intended Recipient, etc.
Organization	Assigned Entity[Organisation], Assigned Custodian, etc.
Entities:	
Device	Device
Location	Health Care Facility, Organisation, Participant
Substance	Labelled Drug, Material, Participant
Workflow Management:	
Encounter	Encounter, encompassing Encounter
Alert	Observation, Act
Supply	Supply
Order	Order, potentially any act in a future mood (RQO, INT, etc.).
Documents:	
Composition	Clinical Document (assuming in a Bundle)

The composition of the resources establishes the document contents (type, date, author, references to the patient) and contains the section hierarchy with resources that make up the structure of the document. FHIR currently fills a new niche using Representational State Transfer (REST) to support Web and Mobile Applications and also has the potential to replace CDA. FHIR documents cover the same functionality as CDA: resources (comparable to sections in CDA) contain both text and software processable data. A FHIR document is a composition of section-level resources.

4.2.5 *REST service*

FHIR is based on REST, an approach to systems architecture. REST provides a set of architectural constraints that emphasises scalability of component interactions, generality of interfaces, and independent deployment of components to reduce interaction latency and encapsulate legacy systems (Fielding 2000). REST is an Architectural style that describes a specific set of web-services that uses a connection-oriented and stateless protocol. In particular, it describes the interaction constraints for distributed systems and is suitable for telemetry devices.

The mobile applications aim at effective management of patients' electronic personal health records via an easy to use interface which can be accessed by multiple actors, namely the patient, the doctor and the pharmacist. FHIR has the implementation benefits to integrate the mobile client and the application's server side and also provide insights into the access control mechanisms that should govern such applications. FHIR offers tools for developers to assemble and present smaller data elements to enhance the context or meaning of the information. An application might be created to remind patients to take certain medications at the right time, for instance. The open Internet standards utilised by FHIR will make a personal health care "account" work much like any other secure application.

RESTful interfaces are identified by the following features:

1. integration is achieved by using well-known HTTP methods;
2. no session information is kept on the server side (stateless system);
3. each resource is mapped to a directory-structure Uniform Resource Identifier (URI);
4. when transferring a resource representation, the data format is either XML or JSON based.

RESTful interfaces are more lightweight by design, comparing to Subjective, Objective, Assessment or Plan technologies (Mulligan and Gracanin 2009) and that is a reason to adopt them in mobile healthcare applications where processing capabilities are limited. RESTful services provide a methodology to create e-Health Application Layer using software architectural styles for Internet of Things (IoT) applications. All REST constraints must apply to a server based on Resource Oriented Architecture (ROA) (Richardson and Ruby 2007). The system should be secured using basic authentication and offer XML and HTML responses. IoT technology platforms can demonstrate light-weight applications developed using a predictable architecture, consistent API specifications and FHIR.

4.3 INTEGRATING THE HEALTHCARE ENTERPRISE

An HIS at the level of a health care facility provides the following functionalities: registration of patients; management of contacts and referrals; reporting; claiming and invoicing; financial accounting and book-keeping. This functionality is provided by means of IT supports for better patient care, as well as standards and APIs in order to achieve integration of applications and interoperate with other Providers. Integrating the Healthcare Enterprise has been successful in enabling interoperability. In order to respond effectively to interoperability you need to set out an Integration Roadmap, showing how it will use open APIs and standards to migrate towards the EHR.

4.3.1 *Registration of patients*

This functionality ensures that all patients (or generally entitled citizens) are registered and their demographic and administrative data are entered and their entitlement to coverage of health care rendered is verified. The Social Security Number (SSN) can be used for verification of entitlement to coverage of health care. Each facility can compare a patient's SSN with the Social Administration's (SA) electronic information to determine if the SSN is valid. When there is a match with the SA database, the correct number is added to the patient's registration record.

4.3.2 *Management of contacts*

This functionality encompasses scheduling of contacts (visits, admissions) and recording relevant data on each contact. All other moves of the patients through the health care facility are recorded as well—e.g. discharges or transfers for specialised services. Each diagnosis and treatment event is recorded, including test results, observations, procedures, medications, clinical notes and outcome. All data on patients, contacts and medical events are stored in one patient's database and can be shared easily in pace with the access constraints within a single or a group of health care facilities.

4.3.3 *Management of referrals*

The functionality represents automatic support of workflow within the health care facilities. Each referral including relevant data is automatically routed to respective destination. It can be retrieved from a queue of tasks (sent referrals), scheduled for accomplishment and the results can be sent back to the sender. The sender can read the results, assess them and include them into the patient's medical history. The referrals can be in-house, to other health care facilities or to other contracted providers. The functionality also includes: tracking of all pending referrals; retrieving status of a single referral; retrieval data from the referral; link to other functionalities, such as scheduling of care or management of contacts; automated support for routine referrals; statistical reporting on "traffic of referrals".

4.3.4 *Reporting*

The functionality encompasses both external and internal reporting. The external reporting concerns producing reports in the specified format (electronically or paper-based) for: national health statistics; public purchaser; public health institution; other health care facilities (e.g. report on hospital discharge). The reports for internal purpose provide data on internal activities relating to frequencies and costs from different perspectives. External reporting relies on a fixed set of predefined reports whereas internal reporting encompasses both a fixed set of predefined reports and report builder tools for generating "ad hoc" reports and analytic tools (so called OLAP tools) for more detailed analysis. Production of a consolidated medical history for individual patients can be regarded as a predetermined form of internal reporting as well.

4.3.5 *Invoicing and claiming*

The functionality is necessary for collecting the revenue of health care facilities and has to be linked to the functionality for registration of patients and management of contacts. The functionality includes: preparation of insurance claims both for in-patient and out-patient cases; preparation of aggregate invoices for public purchaser; production of receipts for patients; payment collection of different types (deposit, prepayment, overpayment, partial payment and non-patient related charges); generating data for financial and accounting components; provision of statistical data on revenues.

Having an interface with patient and performance registration, the system facilitates obtaining accurate demographic records of the patients and calculates the charges automatically according to the patient's attendance information, such as patient status and services consumed. The system has to provide contract control by calculating the cumulative outstanding charges and periodically issuing bills to the discharged patients. The accounting and statistic offices can benefit from the reports regularly generated from the system to monitor the billing statuses of the hospitals and to take any follow-up actions on the debtors if necessary. The statistical reports from the system allow the account and statistical office to perform financial analysis.

4.3.6 *Financial and cost accounting*

This functionality enables to keep and present a consolidated financial position of the health care institution. Actually, this is a subset of the Enterprise Resource Planning functionality. The functions to be supported by the system are: financial control and planning; chart of accounts for analysis and performance comparison; payment service tools; tools for flexible financial management reporting; availability for effective budget control on all budget levels; performance and expenditure planning; production of financial rapports for health care and financial authorities.

The objective of cost accounting is to manage and monitor the effectiveness and efficiency of health care services delivered to the patients. By means of this functionality it is possible to calculate the total costs of the institution and allocate them to services (service unit costs). Based on cost accounting, the system must trace the cost of the services to patient files. The functionality has to enable to calculate and trace the unit costs of services. It interfaces with the registration of the rendered medical services, statistics derived from financial (book-keeping) and management accounting for analysing the performance. Financing and Management information need: a) to record revenue, expenses, and statistics; b) to analyse costs per service, per case and per patient treated; c) to support reporting. The information model shall determine the result of production as a combination of services provided by various units that provide intermediary services.

4.3.7 *Interoperability of e-health organisations*

Interoperability is a requirement for the successful deployment of EHR by enabling access to all relevant information at the diagnostic decision moment, regardless of location. In order to offer easier access to healthcare information EHRs are supporting the integration of heterogeneous HISs and interoperability of e-Health organisations. It is a system resulting from the cooperation of several subsystems that need to successfully exchange information relative to a specific healthcare process.

There is a challenge for the healthcare industry in achieving interoperability among applications provided by different vendors. Each department or clinic may use multiple applications to share clinical and administrative information among applications. Unfortunately, each application may support multiple communications interfaces that must be continually modified and maintained. Achieving interoperability among heterogeneous Healthcare Information Systems (HIS) is important because it will reduce costs associated with healthcare and contribute to more effective and efficient patient treatment, management, and care.

We addressed the major technical challenges for enabling interoperability between HISs. The proposed approach capacitates the integration of a wide range of heterogeneous data sources and also supports the decision making process through its OLAP and data-mining facilities. It builds a suitable solution for integration of high volumes of complex healthcare data. Since the data of participating healthcare organisations are virtually tied in a network and not physically replicated into a central data store, the solution incorporates possible security threats for the sensitive patient data. This approach is a step towards decentralisation of security assurance and towards strengthening of interoperability in the healthcare sector, which results in a better service for the patient, the medical personnel and the administrative staff.

4.4 TECHNOLOGY INTEGRATION OF THE INFORMATION RESOURCES

To solve the problem of semantic interoperability between subjects of an HIS, a number of standards are used: systematised nomenclature of medical terms, a range of laboratory and

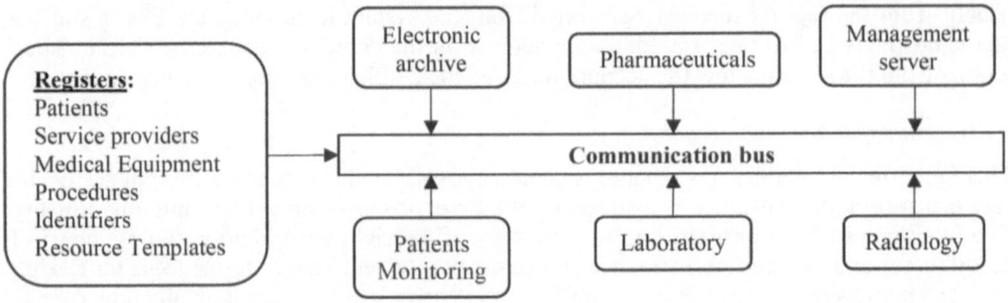

Figure 4.1. Infrastructure of the components for determining EMR.

clinical studies, clinical terminology treatments, medical classification of diseases, medications normalised names and HL7 syntax standards or exchanging DICOM image diagnosis data. The basis for application integration is the middleware software that provides message delivery using the communication bus. The management server implements the format of documents and controls the routing. Fig. 4.1 shows the general components which form the medical records when providing patient services in several domains of the healthcare.

Typically, MIS (medical information system) forms data sets to speed up processing. Transferring a data set is appropriate for medical statistics or payment for treatment. But if you need to take decisions regarding patient care, you must send a whole document to determine the semantics of the messages. Requirement standardisation should be focused on the functionality of the software to create and send messages to fill EHR (Electronic Health Record) "container" by a set of EMRs (electronic medical records). Information support of EHR consists of patient registry, EMR repository and EHR registry. The register includes patient identifiers to support EMR connections at medical organisations with the EHR register. Personalized EMR are formed on the basis of relevant templates while EHR on the basis of EMR and logical connections between patient and entries (visit the patient in case of medical care, hospitalisation, and so on).

For compatibility of information systems the following components are applied: parsing message and storing data; access to information through the Web Services for interaction with external HISs; audit when dealing with personal data of patients and authorisation of access to data; integration of departmental a HIS with insurance funds. The information exchange system using the XML language provides message invariance regarding the type of software. The main advantage of XML is that this standard transports metadata along with the relevant data.

4.4.1 *EMR database*

The data model is defined by relations between the classes of objects and dimension attributes of entities. When designing a database, one should consider clinical terms, data types and attributes binding EMR to the patient. The design of the EMR repository should implement a scheme with a fixed number of database tables. That is, the variability range of medical procedures should lead only to an increase in terminological dictionaries, and not a change in the physical database schema.

The need for storage of XML-data in a relational database is due to the use of XML for data exchange. The key problem is that relational data is strictly structured while XML data are structured arbitrarily. Each column of the relational table has a data type that can effectively perform their processing. XML can have different data types, making it difficult to process information.

In recent years, several XML data control systems have been proposed, using relational database management system with data management capabilities of the XML format

(Krivonos and Khorozov 2014). There are several methods of storage designed for hierarchical XML documents into the object-relational model. The relational model is the most suitable for practical application, but the dimension attributes and complex data types are consuming design time. In a relational database an XML document is saved as symbolic objects in columns CLOB or BLOB. Disclaimer structure (XML-parsing documents) saves time and increases productivity. The "analyser" for parsing XML-documents is a data retrieval tool. Cost analysis of data query and update increases in proportion to the size of the XMLdocuments. If XML and relational data representation are combined, the appropriate balance between two possibilities must be found: to save documents in the XML column or conduct decomposition (shredding) of data storage in relational format.

To design a database with storage capabilities for XML data it is proposed to use relational approach of the document context and a hybrid approach at semantically coherent document sections. A document section is selected as the logical unit of storage media, like the table records. For each type of sections the XML Schema Definition (XSD) collection is constructed on the database server, according to the template.

For data exchange it is necessary to provide a module for loading data into the DB and XML data format extraction from the DB through a programmatic interface. API XML and ASP Web are common to software environments. To store XMLsections in a DB, objects are loaded into memory and then access to its components is organised through XPath queries. Data extraction is implemented using XPath expressions and indicating the way to the contextual elements using API. For example, you can easily request the section "Results" of the PMR for the examination episode. The following solution for EMR database design is proposed:

1. determining the protocol for information exchange between systems;
2. using the HTTP or TCP/IP transfer protocol;
3. serialisation of XMLfile data and deserialisation of the received message;
4. forming a collection of XMLschema sections;
5. loading the context and XMLsections to the database.

Decomposition of the document formally leads to its transformation. Thus, for the formation of a new message you should complete the assembly. EHR data may be prepared with queries to the EMR database for patient and specific EMR sections.

The conditions for access to patient information in real time are a key aspect of modern medical data processing servers of. The scheme transmission of the medical data is used in HL7 format. The elements of terminological dictionaries and healthcare information resources should be available on the help server. The Healthcare departments exchange data through the Web-service that provides authorised access to the EMR DB. Information support of the patient EHR and access to the health sector resources involves the Cloud technology. For effective EMR sharing and storing there implementation of communication standards is needed; development of standard interfaces (to achieve semantic HIS interoperability) and approval of electronic registration forms by medical institutions and health insurance organisations.

4.4.2 *Internet of things implementation*

The IoT describes a system which gives entities a virtual representation and registers them to a network. The advantages of the IoT are enhanced identification capabilities of things. Since all entities are tagged and have a unique representation, every scanning procedure will give a clear association to a virtual entity and its data. Today's mobile devices, for instance, have similar processors to classic desktop computers, enhanced storage capacity and nearly permanent Internet access. Each has a unique address provided by an Internet service provider (ISP). They fulfil all IoT criteria and are also propagating information. The IoT using REST as a standard for synchronous data exchange actively promotes RESTful web architecture, supporting any network capable device as its extension.

The Entity is a global concept embracing a physical and/or virtual part. An Entity can be composed of a Physical Entity and have a Virtual Entity. An instance of a Physical Entity would be a sensor measuring a physical property. The Virtual Entity stands for the virtual part of an Entity. As such, a Virtual Entity would hold the value of a temperature sensor attached to the Entity. The Virtual Entity reflects the character of the Physical Entity. Fig. 4.2 shows the elements composing the model.

The technological implementation of a system is based on transmission of data through the Internet. The eHealth facilities are equipped with devices for data collecting. In the IoT these devices are interconnected and share data over the network. This renders it possible to design independent operations, ranging from simple alarm system if an analysis shows a critical value for a specific patient's vital signs or life support system. Network topology consists of a set of sensors, a data registration unit, a data communication module and a data collection unit (Fig. 4.3).

Data collected from a node of the sensors are processed by the logger for further transmission. In terms of hardware and software, the logger uses a communication module for data transmission through a web interface on a DB server. For example, monitoring scenarios are supporting several types of indicators. Sensors record: a) numerical-tuple or discrete values (temperature, pulse oximetry, blood pressure) and continuous signal (ECG). The ECG signal detector provides a serial data stream, while the logger determines the PQRST-segment, using an algorithm for analysing ECG signals.

The apparatus may provide local storage of data, which are used for recording signals together with the results of their processing. The data logger together with the measuring

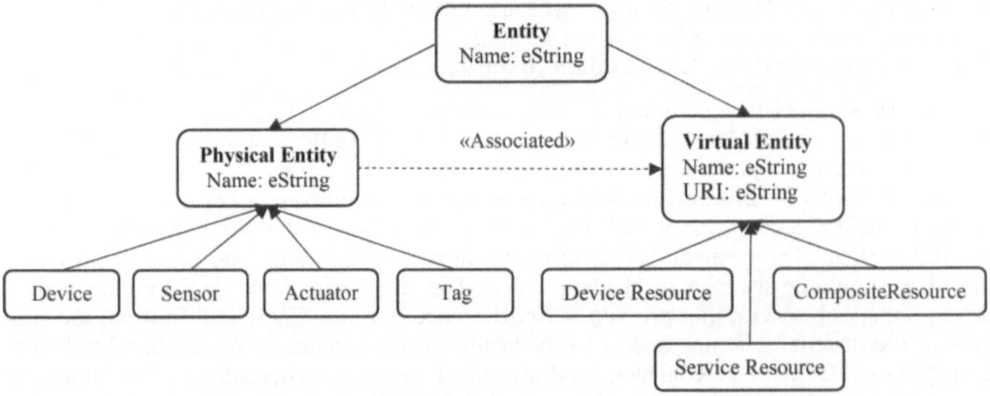

Figure 4.2. Meta-model for the web of things components.

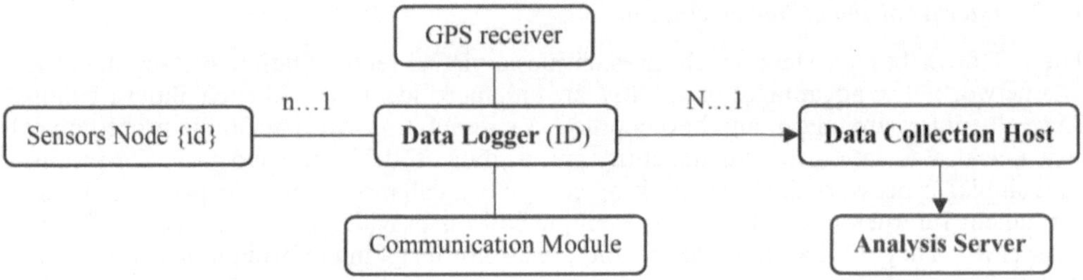

Figure 4.3. Topology of the network.

sensor may be actuators, which implement feedback if necessary. The information flow between the components of model is this: the application sends an XML with the patient ID, the Host reads the XML request, identifies the needed data, and sends it to the Emergency department where the data are read and filled in the chart.

The data acquisition system sends the e-Health files to the Remote Data Collection Host (RDCH). The RDCH has the responsibility to keep the body of information permanently in dedicated spaces. The Client/Server scheme permits control of the monitoring process. The monitoring process is distinguished by a basic and an emergency operational mode. In both modes, the monitoring process requires real-time recording of the individual's vital sign, context, and geolocation data. The recorded data are included within one file and sent to the RDCH.

REST brings services into the browser: resources can be linked and bookmarked and the results are visible with any Web browser to be able to interact with the service. To achieve this, REST proposes two basic rules: 1) the application model is transformed from operation-centric into a data-centric one. This means everything that offers services becomes a resource (e.g. a temperature sensor is a resource of the sensor node resource) that can be identified unambiguously using URIs; 2) the four main operations provided by HTTP (GET, POST, PUT, DELETE) are the only available operations on resources, they define a uniform interface with well-known and shared semantics.

By creating RESTful APIs to integrate the services offered by devices and objects such as wireless sensor networks with any other Web content, a system of RESTful devices can facilitate the creation of cyber-physical mashups.

4.5 EHEALTH RESOURCE SERVER

This section will cover the design process of the FHIR Server resources which are relevant to eHealth. The aims are identifying methods applicable to reduce the effort in building RESTful services for the IoT. The main purpose is to provide a methodology to create web-services for eHealth. In an eHealth environment data consistency is a key element for a reliable service. All REST constraints must apply to a server and it must be based on ROA. The system should be secured using basic authentication and offer at least XML and HTML responses. For eHealth Resource Server generation we are using three technologies: XSD file to describe the resources; Web Application Description Language file to define the functionality based on the XSD; Entity Relationship Model for storage on a database.

The methodology for creation of the resources database has the following steps:

1. Evaluate Data Sets (evaluate medical records to extract the entity sets).
2. Split data sets into resources (the sets are split into compositions and resources).
3. Name the resources with Uniform Resource Identifier (URI schema).
4. Design representations (define messages served by the server to the client and vice-versa).
5. Resource storage (design the entity database relationship model).

The first step in building the service is to define the entities the service offers. For the eHealth Server everything starts with a use case representing a patient and a provider. The paths implement the structure are given by the object diagram in Fig. 4.4.

A URI scheme is represented by the resources and still respects all constrains on addressability. URIs can be separated into authority, path and query. A composition is therefore on a shorter path than a service resource. Queries can be used to filter lists according to their association. The URI scheme is shown in Table 4.2. This is a representation of the URIs for the entities defined in Fig. 4.4.

For each URI it is necessary to define which of the methods GET, POST, PUT or DELETE are applicable (Table 4.3).

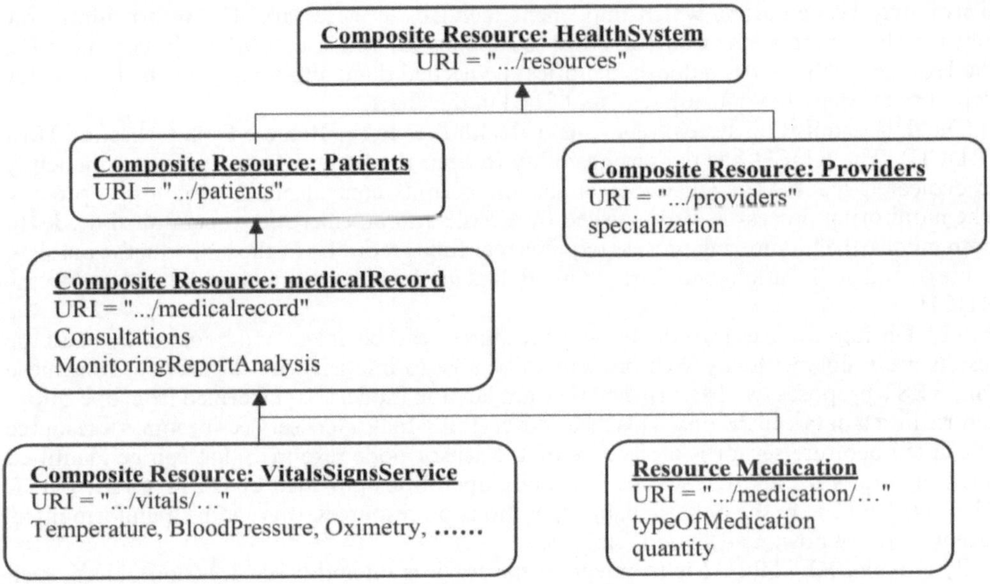

Figure 4.4. Object diagram of the resource.

Table 4.2. URI schema for the eHealth Server.

Authority:	
Server root resource	http://localhost:8080/eHealthServer/resources
Path:	
health system	…/
all patients	.../patients
specific patient	.../patients/{pid}
medical records	.../patients/{pid}/medicalrecord
vital signs	…/patients/{pid}/medicalrecord/vitals
medications	.../patients/{pid}/medicalrecord/medication
Queries:	
caregiver for patients	.../patients?cid = {cid}

Table 4.3. Overview of the HTTP methods used.

Method	URI
Authority:	
GET	http://localhost:8080/eHealthServer/resources
Path:	
GET	…/
GET, POST	.../patients
GET, PUT, DELETE	.../patients/{pid}
GET, PUT	.../patients/{pid}/medicalrecord
GET, PUT	.../patients/{pid}/ medicalrecord/vitals
GET, POST, DELETE	.../patients/{pid}/medicalrecord/medication

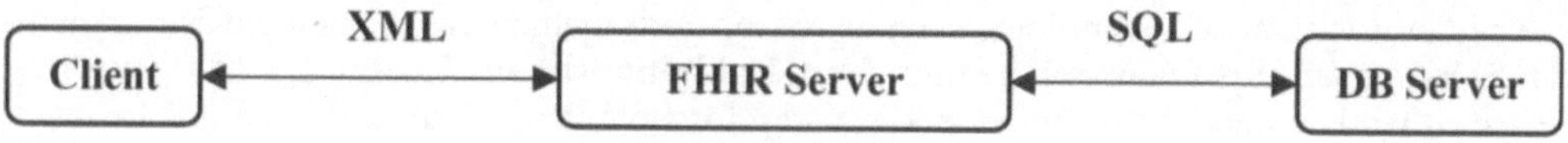

Figure 4.5. FHIR server data transition.

The eHealth Server uses Client for URL (c-URL) in a terminal shell and provides basic manipulation operations such as adding, removing and modifying of a resource. The server implements the transition from request in XML or URI form parameters to database entries and the transition from a database entry to a HTTP response containing an XML body (Fig. 4.5).

The system counts three components: a client, a server and a database. The client communicates to the server using an URI request containing XML-encoded messages. The server transforms those messages into an SQL request and assures transition of data to SQL. The business logic therefore handles both APIs and transition from one to the other. It binds all components together and ensures valid transitions. The system is able to transmit the data to the server and also support bidirectional communication receiving control signals from the server. The eHealth Server is responding in XML format answers as usual. This XML message needs to be designed for automatically marshalling and unmarshalling client responses and has the structure which is defined by an XSD file.

4.6 VITAL SIGN REPRESENTATION FRAMEWORK

Consider a framework for a patient's vital signs representations in terms of XML schema diagrams. We use an XML schema to define and specify the vital signs. Vital signs represented in accordance with this framework are suitable for forwarding, processing and rendering within a heterogeneous environment which applies different formatting standards or uses a diversity of communication technologies. This section is about the design of a data format for e-health record and illustrates how to construct the data payloads protocol for exchange of vital signs between remote peer entities. E-health applications have been receiving more and more attention due to their ability to reshape healthcare delivery, like continuous and ubiquitous tele-monitoring and tele-treatment. The model for (encoded) vital sign representations and conversions enables the specifications of data that satisfy the requirements of the end users (healthcare professionals) and match with the capabilities of the RESTful platforms.

The telemonitoring system includes the following hardware and software components (Costin et al., 2010). A wireless biophysical and biochemical sensor network sub-system, composed by transducers for vital signs and data transmission. The platform of heterogeneous sensors is a key functional feature of the telemonitoring system. The devices allowing monitoring of the vital parameters are the following: a) the body temperature module; b) the blood pressure module; c) the oxygen saturation (SpO2) module that also computes the cardiac rhythm; d) the respiration module; e) n-leads ECG module records and transmits data through a radio transceiver interface.

Application Software is designed for:

1. Data acquisition and processing, signal analysis, data transmission to server;
2. Interface programs for duplex communication between sub-system and central server;
3. Alarm generating and transmission to the Ambulance; here necessary decisions are taken, based on specific medical knowledge;
4. Extra data processing programs at the central server level (medical statistics, research informatics management, etc.).

The development of technology, such as sensor nodes which are replacing existing medical devices, opens up a new opportunity for eHealth innovation. The linkage of monitoring devices directly to healthcare providers is a step forward for preventative health. The interoperability across multiple platforms and devices together with the integration of data into efficacious resources may affect clinical decision support. Mobile health gateways will emerge to facilitate connection and transfer from mobile networks to existing HISs.

4.6.1 *Specification of the framework*

We are using patient-centric approach to design the framework. This makes a framework which support diverse vital sign representations for a particular patient. The XML schema diagram can visualise the hierarchical structure of the PMR. The structure is present in Fig. 4.4. It consists of two parts: Patient Demographics Records (PDR) and Vital Sign Records (VSR). A patient's ID item of PDR is the mandatory element among all the child-items.

This is to fulfil the privacy requirement of the end-user stakeholder. In addition to conventional sensors for Vitals, a Global Positioning System (GPS) module is used for user geolocation. Based on the medical scenarios, a few vital sign types are supported, as shown in Fig. 4.6. These types could be divided into the following categories: a) the numerical-tuple vital sign type, e.g. temperature, SaO2 and blood pressure; b) the waveform vital signs type, e.g. ECG.

Each vital sign record has a Data section and a Property section. The property section is specified as a complex type. It provides the necessary information for processing the vital sign data correctly.

The data section representation of the Vital signs is shown in Fig. 4.7:

For example, a recording of measuring the heart rate and blood oximetry in XML-format is represented in Fig. 4.8:

The package data is composed of a set of discrete values of blood oxygen saturation and pulse rate. Monitoring of the patient data record is delivered together with the results of preliminary signal processing. Some of a patient's vital signs data range is given in Table 4.4. The following physiological conditions cause alerts when the deviation from the norm exceeds the specified range:

The Host Analytical Server, which accepts the vital data, takes into account the critical values of the vitals to start the trigger procedure for an ambulance. The nodes of the network consist of transducers and actuators. The Server of the analytical system, taking into account

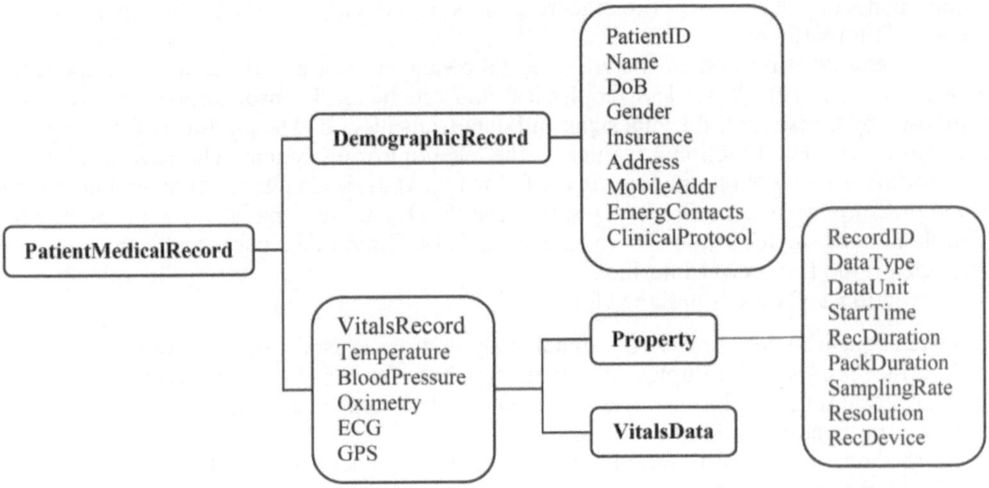

Figure 4.6. Vital signs for patient medical record.

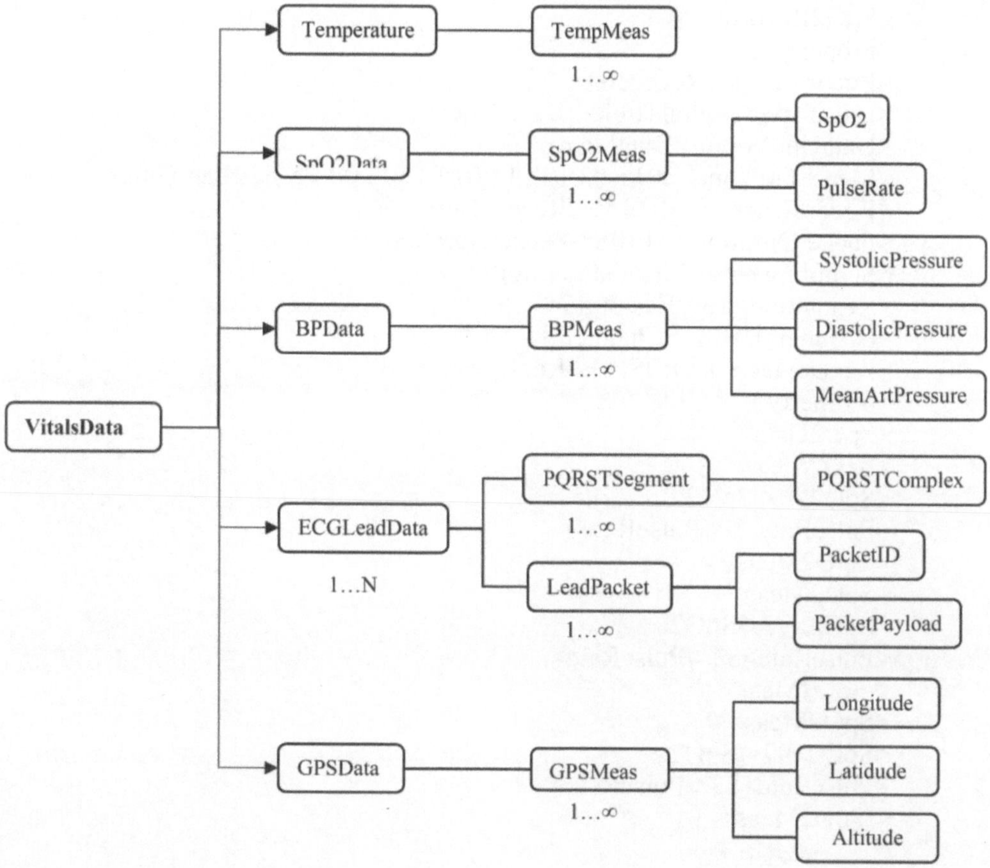

Figure 4.7. The vitals data complex.

identifiers of the controller board, allows realising activation of procedures associated with the provision of emergency medical care from the actuator which plays the role of a syringe with an injection.

4.6.2 *Database for telemonitoring system*

Wireless devices can transmit remote diagnosis via the Internet network for further analysis of the information and provision of feedback. The main problem of data acquisition is the parsing and analysis of messages before loading information into the database. A technology platform to provide a unified application interface scheme for data analysis regardless of the structure of signals is proposed. The software adapter is used to bridge the gap between the interface device and the server platform.

It must be taken into account that recording medical data in waveform and discrete form that affects the structure of the data warehouse is needed: the identification of the patient; identification of the type of sensor spatial coordinates; date and time of measurement parameters. The DB for PMR should combine different types of monitoring data. This eliminates the need to restructure the data warehouse to add new types of sensors. The database is needed to take into account the spatial coordinates, date and time of measurement and the values of each subscriber network. The structure of the medical record database is described by the following diagram (Fig. 4.9):

```
<SpO2Record>
<Property>
<RecordID>1</RecordID>
<DataType>DigitalTuple</DataType>
<DataUnit>bpm</DataUnit>
<MeasStartTime>2015-03-01T14:01:01+02:00</MeasStartTime>
<RecordDuration>PT1M</RecordDuration>
<PacketDuration>PT10S</PacketDuration>
<SamplingRate>256</SamplingRate>
<Resolution>16</Resolution>
<Channel>CH02</Channel>
<RecDevice>DeviceSPO</RecDevice>
</Property>
<SpO2Data>
<SpO2Meas>
<SpO2>0</SpO2>
<PulseRate>0</PulseRate>
</SpO2Meas>
<SpO2Meas>
<SpO2>97</SpO2>
<PulseRate>83</PulseRate>
</SpO2Meas>
<SpO2Meas>
<SpO2>97</SpO2>
<PulseRate>82</PulseRate>
</SpO2Meas>
....................
</SpO2Data>
</SpO2Record>
```


Figure 4.8. The SpO2 vital sign record.

Table 4.4. Data for determining the critical indicators.

Indicators	Data range	Norm
Body temperature (T)	35,6–42,2 C	36,6 > +/– 11%
Pulse rate (PR)	30–200 (bpm)	60–90 > +/– 11%
Oxygen saturation (SPO2)	90–100%	95–98 > +/– 11%
Systolic pressure (SP)	40–250 (mmHg)	120 > +/– 11%
Diastolic pressure (DP)	15–200 (mmHg)	80 > +/– 11%
	Calculation	
Mean arterial pressure	(SP + 2 * DP) / 3	(90–100) mmHg
Pressure Index	(SP-DP) / SP	< 0,5 (50%)

To create service-oriented applications the software platform Windows Communication Foundation (WCF) is used, which allows you to combine logic delivery, storage and analysis of the data. WCF services are forwarding, processing and storing data on the Server.

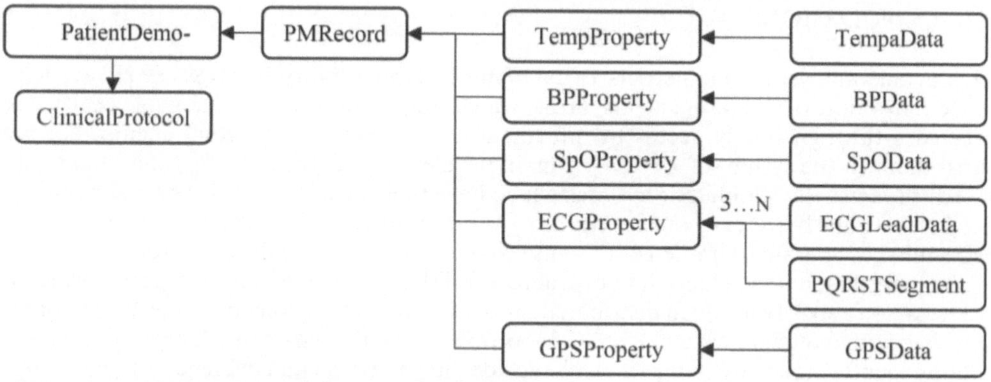

Figure 4.9. Block diagram of the database fragment.

	docID	LeadedDateTime	rootID
1	1	2014-02-12 14:22:23.000	907380D9-9888-47F4-A6F3-7956384C4311

	pdID	docID	PatientID	DateofBirth	Gender	Race	Height	Weight	PatientLastName	PatientFirstName
1	1	1	PC-123	1953-12-07	Male	Ukrainian	180	90	Khorozov	Oleg

	mdID	docID	VentricularRate	AtrialRate	PRInterval	QRSDuration	QTInterval	QTCorrected	PAxis	RAxis	TAxis	QRSCount	QOnset	QOffset	POnset	POffset	TOffset	ECGSampleBase	ECGSampleExponent
1	1	1		NULL	148	120	420	443	44	-61	86	NULL	217	277	143	194	427	500	0

	LeadByteCoun...	LeadSampl...	LeadAmplitudeUn...	LeadAmplitudeUnits	LeadHighLimit	LeadLowLimit	LeadID	LeadSampleSize	LeadDataCRC32	WaveFomData
1	10000	5000	4.8800000000	MICROVOLTS	2147483647	268435456	I	2	3994716513	//////////////7////+//7//v/9//z//P/8//3//v/9//v/+//7//z//P/8//3//A...
2	10000	5000	4.8800000000	MICROVOLTS	2147483647	268435456	II	2	624503064	A/9//3/A/9//3/A/9//3//P/9//3/A/9//z/+v/7//r/+4/5//n/+4/6//n/+v/...
3	10000	5000	4.8800000000	MICROVOLTS	2147483647	268435456	V1	2	1873944278	FgAWABYAFgAWABUAFQAVABUAEwATABMAFQAUABQAEgASABIA...
4	10000	5000	4.8800000000	MICROVOLTS	2147483647	268435456	V2	2	3731892261	HAAbABoAGQAYABYAFQAVABUAFQAUABQAFQAVABYAFAATABMA...
5	10000	5000	4.8800000000	MICROVOLTS	2147483647	268435456	V3	2	1384517557	FAAUABQAFAATABIAEgAQABAAEAARABEAEQAQABAADwAPAA8AE...
6	10000	5000	4.8800000000	MICROVOLTS	2147483647	268435456	V4	2	2233107026	DgAOAA4ADgAOAAwACwAMAAwADAAMAAwADAAMAA0ADAALAAsA...
7	10000	5000	4.8800000000	MICROVOLTS	2147483647	268435456	V5	2	4084894931	CwALAAsACwALAAsACwAKAAoACgAJAAgACAAHAAcABwAHAAcACA...
8	10000	5000	4.8800000000	MICROVOLTS	2147483647	268435456	V6	2	3703775523	/P/9//7///8AAAAA////////A/9//3/A/9//7//v/+//7//v//////////////// ...

Figure 4.10. ECG record to DB Tables.

The Code procedure on SQL server implements the logic of parsing and loading data into the database from XML files. To load the data it is necessary to parse an XML file on the elements and export data to relational tables of the SQL server. Apply the standard algorithm for loading XML data to SQL server: a) record control code that implements the logic of parsing an XML file; b) create the loading procedure of data in the database table. After transferring the measurement results through Web-service, files are checked for validity and loaded to the database. The example query for analytical parameters and ECG waveform to a relational database are presented in Fig. 4.10.

When applying the reverse procedure, query forms ECG.xml file with the values of the elements that were initially loaded to database.

The system provides two levels of information processing. At the first level the critical values of the physiological parameters are detected, the result report is formed and transmitted to the analytical centre. At the second level the patient's data are recorded in the form of medical reports. At the same time the patient's medical reports can be accessible via Web-services for healthcare providers which should be notified for making appropriate decisions. The patient's critical vital signs are transmitted to the monitoring center for collection and processing.

4.7 CONCLUSIONS

This article analyses some aspects of enabling interoperability in HISs. EMR are supported by multiple heterogeneous HISs which improves the efficiency of healthcare processes through timely access to information. EMR consist of components that are subsystems to the system of EHR. The main problem comes from the fact that those subsystems are usually standalone systems which have their own functions. These subsystems need to act jointly to serve the functions of the overall system. Therefore, heterogeneous HISs must cooperate to implement the use cases of a specific healthcare process.

The impediments to achieve interoperability in HISs are divided into categories, which it is necessary to specify for open distributed processing systems: a) Technical, relating to problems in communication, which include connection, syntactic, data consistency; b) Semantic, relating to information exchanged, which include interpretation and reference; c) Functional, relating to the difference between the expected behaviours of a component, which include functional model, embedding and intention; and d) Logistical, relating to the impact of the design on the system.

When defining the system requirements, designing or testing, impediments need to be addressed to ensure that the resulting system will successfully support the healthcare processes. We presented a mobile application that leverages the emerging e-Health standard HL7 FHIR, which enables easy implementation and integration of healthcare systems, using technologies that are especially suitable for mobile environments. The application, based on the information resources of the FHIR standard, facilitates the interaction between healthcare professionals and patients and offers the means for continuous monitoring of their health. An essential step is to enable efficient sharing of health measurement streams via FHIR resources. The HL7 FHIR standardisation initiative builds on the previous HL7 standards and allows easy implementation.

The standard defines lightweight resources that function as an interoperability base between communicating healthcare systems, while supporting resource extensibility and use of customised vocabularies and ontologies. On the client side, each operation of the REST interface can be easily bound to specific components. Firstly, it preserves the responsiveness of the application while fetching the relevant resources. Secondly, it supports on demand resource consumption and aggregation. On the server side, the REST interface facilitates the easy enforcement of the access control policies, by developing an HTTP filter handling all incoming REST calls.

Moreover, FHIR resources and their relations are easily mapped to a relational database schema enabling their storage and retrieval. The REST approach to build Health services on the Web originates by the necessity of the evolving the health service to accommodate emerging needs that are complex. The architecture inherits some of the basic Web techniques to implement a highly scalable and flexible system that is able to satisfy the current needs. Without changing the IT infrastructure, this approach could be implemented by an interface to access data and resources.

REFERENCES

Clinical Data Standards in Health Care: Five Case Studies, 2005, from: http://www.chcf.org/publications/2005/07/clinical-data-standards-in-health-care-five-case-studies.

Costin H. et al., 2010: Real Time Telemonitoring of Medical Vital Signs, Proceedings of the 2nd WSEAS International Conference on BIOMEDICAL ELECTRONICS and BIOMEDICAL INFORMATICS, ISSN: 1790–5125, ISBN: 978-960-474-110-6, pp. 127–135.

Fielding R., 2000: Architectural Styles and the Design of Network-based Software Architectures. PhD thesis, University of California, Irvine, USA.

Health Level 7 XML—http://www.hl7.org.

HIMSS 2013: http://www.himss.org/library/interoperability-standards/what-is-interoperabil.

Khorozov O.A., 2014: **Формування бази даних електронних медичних записів** (in Ukrainian, 'Database of Electronic Medical Records Formation'), " ***Компьютерная математика*** ('*Computer Mathematics*'), 2014, №1, pp. 61–68. [URI: http://dspace.nbuv.gov.ua/handle/123456789/84810].
Krivonos Y., Khorozov O., 2014: Heterogeneous Medical Information System Integration, Journal of Automation and Information Science, v. 46, i11, pp. 1–8.
Mulligan G., and Gracanin D., 2009: "A comparison of SOAP and REST implementations of a service based interaction independence middleware framework", in *Proceedings of Winter Simulation Conference (WSC)*, pp. 1423–1432.
Richardson L., Ruby S., 2007: *RESTful Web Services*. O'Reilly, 2007.

APPENDIX A. HEALTHCARE INTEROPERABILITY STANDARDS

Interoperability depends upon syntax and semantics. A number of standards have been developed to address the requirements of both types of interoperability. The following categories of standards are used in healthcare:

Messaging standards – messaging standards outline the structure, content and data requirements of electronic messages to enable the effective sharing of information. The term 'message' refers to a unit of information that is sent from one system to another, such as between a laboratory information system and a GP's clinical information system. Examples of messaging standards are HL7 v2.x for administrative data and Digital Imaging and Communications in Medicine (DICOM) for radiology images.

Terminology standards – terminology standards provide specific codes for terminologies and classifications for clinical concepts such as diseases and medications. Terminology systems assign a unique code or value to a specific disease or entity, for example, the ICPC-2 code for 'asthma' is R96. Terminologies are used primarily to capture clinical information at the point of care. It is necessary to select different classifications and terminologies in combination to enable complete coverage across all of healthcare. Examples of terminology standards include ICPC-2, ICD-10, SNOMED CT for clinical terms and LOINC for laboratory results.

Document standards – document standards indicate the type of information included in a document and also the location of the information. Examples of document standards include the paper-based SOAP standard and CDA for electronic sharing of clinical documents. HL7 have developed document-standard specifications for a continuity of care document (HL7 CCD) and a discharge summary (HL7 DS).

Conceptual standards – conceptual standards allow the transmission of information between systems without any loss of the meaning or context of that information. For example, the HL7 Reference Information Model (RIM) provides a framework for describing health information and the context around it, i.e. who, what, when, where and how.

Application standards – application standards determine the implementation of business rules for software systems to interact with each other. For example, application standards can allow a single user to log in to multiple information systems in one environment, allowing efficient access to the required health information. This can facilitate the simultaneous viewing of health information across multiple databases that are not electronically integrated.

Architecture standards – architecture standards define a generic model for health information systems. They allow the integration of health information systems by providing guidance to aid the planning and design of new systems as well as the integration of existing systems. This is achieved by defining common data elements and business logic between systems. For example, the CEN standard ENV12967 Healthcare Information Systems Architecture provides an open architecture that is independent of technical specifications and applications.

CHAPTER 5

Independent devices and wireless sensor networks for agriculture and ecological monitoring

Y.G. Kryvonos, V.O. Romanov & I.B. Galelyuka
V.M. Glushov Institute of Cybernetics of NAS of Ukraine, Kyiv, Ukraine

W. Wójcik & T. Zyska
Lublin University of Technology, Lublin, Poland

E. Amirgaliyev
Institute of Information and Computational Technologies, Almaty, Kazakhstan

INTRODUCTION

Plants as well as all organic matter suffer due to impact of stress factors of natural or anthropogenic origin. Using of smart portable device to diagnose the plants state under impact of such stresses to agriculture and environmental monitoring enables us to apply advanced measures in time to protect the plants from diseases, to save and increase crop capability and reduce the expenditure on unit of product and harmful substances content in plants. Application of this kind of the devices is very important in precise farming, since this new agriculture technology needs the automation of all of its elements.

Unforeseen climate changes and difficulties in using of statistical data demonstrated huge necessity to develop models for forecasting crop yield and influence of climate on it. These models have to be based on current, but not only statistical data.. Input data for these models is urgent and important task. To get such data, the Earth remote monitoring techniques (e.g. space observations) and acquiring data from surface tools for monitoring of plant state can be taken into consideration. It is important to mention that these data are very significant for precise agriculture technology, which also operates on models for forecasting crop yield and is used to minimize the costs (e.g. water, fertilizers etc.) and maximize the harvest.

Therefore, acquiring of current and objective data of plant cover state is very important factor in most cases, which creates a future strategy of maintenance of agricultural lands, woodlands, parklands and proper decision making. Certainly, it would be ideal to obtain information about improvement or worsening of plant cover state beforehand, not after the event. It lets to avoid increasing costs and save harvest and green plantations from a possible loss.

The plants ability to complete photosynthesis depends on a range of factors including stresses caused by environmental conditions. Absorbed light energy used for photosynthesis is dissipated by emission of heat and re-emission of small but diagnostically significant amounts of absorbed radiation such as long-wave red/far-red light energy. This re-emission of light is named as induction of fluorescence of chlorophyll. Maximum induction of fluorescence of chlorophyll occurs in the red region of the spectrum (685 nm) and extends into the infra-red region to around 800 nm. Chlorophyll fluorescence emission from the whole leaf systems is too weak to be observed with the naked eye, but it can well be observed with special devices such as fluorometers.

As a result of the program of Presidium of National Academy of Sciences of Ukraine (NASU) "Development in the field of sensor systems and technology" in V.M. Glushkov Institute of Cybernetics of NASU, portable smart devices of "Floratest" family aimed to diagnose the influence of stress factors on the plant state has been developed (Romanov et al. 2007). The portable device measures Chlorophyll Fluorescence Induction (CFI) without the plant destruction. Using the curve of CFI (alike the cardiogram) allows to diagnose an influence of one or more factors on the plant state.

5.1 STATE OF THE ART

The method of chlorophyll fluorescent induction has been known for more than 80 years (Kautsky and Hirsch 1931). This is a sensitive, non-invasive method of prompt estimation of plant state. The matter of the method is following: if you illuminate chlorophyll by blue light, it begins to emit light in red spectral region. It is possible to estimate the plant state by the curve form of intensity of emitting light of chlorophyll. But to detect an unique relation between the plant state and the intensity of emitted light of chlorophyll, it is necessary to develop special methods, which allow detection the type of stress factors (for example, drought, freeze etc.), natural of anthropogenic factors, kind of viral infection or bacterial load, sort of plant and so on. It should be mentioned that for the different sorts of the same plant there are many different kinds of of such a relation.

The method of chlorophyll fluorescent induction has been studied and developed in well-known universities of the USA (R.B. Peterson and D.E. Aylor, Department of Biochemistry and Genetics (R.B.P.) and Department of Plant Pathology and Ecology (D.E.A.), The Connecticut Agricultural Experiment Station, 123 Huntington Street, New Haven, Connecticut 06511, Gray DW, Cardon ZG, Lewis LA. Department of Ecology and Evolutionary Biology, University of Connecticut, Storrs, CT 06269–3043, Zhu, Xin-Guang, Govindjee, DE Sturler, Eric, Long, Stephen, University of Illinois, Baker, Neil, University of Essex), Japan (Isamu Ikegami and Sakae Katoh, Department of Biophysics and Biochemistry, Faculty of Science, University of Tokyo), Canada (Stephen S. Bates. Department of Fisheries and Oceans, Marine Ecology Laboratory, Bedford Institute of Oceanography Dartmouth, Nova Scotia, Canada), Western Europe (Alfred R. Holzwarth, Max-Plank Institute, Germany) and Russia (Moscow state university) and etc.

Precise farming is based on control of productivity of agricultural manufacture with taking into account changing of performance of external environment. According to precise farming agriculture, fields are digitized and remote monitoring is done with systematic ground measuring of plant states to make a final decision.

The aim of precise farming is to get maximum crop with optimum expenditure on fertilizers, herbicides, biology active additions, including energy and water recourses.

Nowadays, the integration of different sensors in networks for data acquisition is used in many areas of human activities. The technology of sensor networks has a promising future. Although many leading R&D organizations and universities in the world are interested in this technology, a lot of bottlenecks remain, for example, sensor resistance to climate changes, reducing of power consumption, possibility to cover large territories, sensor reliability, repeating of measure results, self-organization of sensors in case of damaging of a separate sensor from the network etc.

Wireless Sensor Network (WSN) is a technology with a promising future. Nowadays, wireless technologies are used in many areas of precise farming and ecological monitoring: spatial data collection, irrigation, variable-rate technology and providing data for farmers, plants and trees protection etc. In fact, a number of studies have been published in the last fifteen years on applications of using of wireless sensor networks in precise farming and ecological monitoring.

Sensor networks for acquiring of data, particularly wireless sensor networks for precise farming, environmental protection and ecological monitoring, now are studied and developed by many well-known R&D organization and universities, for example, Department of Environmental Sciences, University of Virginia, USA (John Porter), Freie Universität Berlin, Institute of

Computer Science, Germany (Tomasz Naumowicz, Jochen Schiller), Taiwan Forestry Research Institute, Taiwan (Chau-Chin Lin), Electrical and Computer Engineering Department, University of Limerick, Ireland (Essa Jafer), Commonwealth Scientific and Industrial Research Organisation, Australia (Chris Knight, Kate Cavanagh, Christopher Munnings), Graduate School of Bioresources, Japan (Takaharu Kameoka, Atsushi Hashimoto) and many others.

Authors of papers (López et al. 2009) and (Haldar et al. 2011) described wireless sensor networks, which are able to acquire data about soil and environment parameters. The main purpose of these systems is monitoring large territories. In (Gomide et al. 2001), mobile field data acquisition system is described, developed to collect data for crop management and spatial-variability studies. Data acquiring and control systems are placed on farm machines. The system is able to conduct local field survey and to collect data of ground water availability, soil compaction, soil fertility, biomass yield, leaf area index, leaf temperature, leaf chlorophyll content, plant water status, local climate data, insect-disease-weed infestation, grain yield etc. Paper (Lee et al. 2002) describes silage yield mapping system, which included a GPS, load cells, a moisture sensor and a Bluetooth wireless communication module.

The moisture sensor and the Bluetooth transmitter were installed on the chopper. Authors of (Lee et al. 2002) had developed a wireless, infrared thermometer system for in-field data collection. The system consists of infrared sensors, programmable logic controllers and low power radio transceivers to collect data in the field and transmit it to a remote receiver outside the field. In (Evans and Bergman 2007), wireless sensors has been used in the system to assist irrigation scheduling using combined on-site weather data, remotely acquired data and grower preferences.

Paper (Camilli et al. 2007) describes a simulation to demonstrate the utility of a wireless sensor network in precise agriculture. Authors of paper (Pierce and Elliot 2008) described a platform that they developed for local and regional networks and implementations of them in two precise agriculture applications. In (Morais et al. 2008), a prototype developed to be deployed on vineyards to monitor crops of this type has been described. Detailed analysis of WSNs applications in medicine and health care is given in (Jeong Gil Ko et al. 2010).

It should be mentioned that in all examined papers, proposed systems measure different parameters of soil, plants and environment (temperature, humidity, biochemical property of plants, even sound range etc.). But none of the systems measure parameters of photosynthesis—the main process of plant vital activity. If it is necessary to estimate plant state and influence of stress factors of natural or anthropogenic origin on them, the most effective and informative way for accurate estimating of plant state is to measure the photosynthesis parameters in plants.

5.2 PRINCIPLES OF DEVICE OPERATION

As a result of external influence, different objects, including biological ones, can emit great amounts of radiation that is not related to these objects' temperature.

All the types of radiation, which were caused by some external sources of energy, are called luminescence. Duration of luminescence after external influence stopping exceeds period of light fluctuations. Luminescence is conditioned by fluctuations of relatively small number of atoms or molecules of substance that become excited under energy source activity. Radiation is a result of transformation of atoms' or molecules' states into fundamental (unexcited) or less excited (they have less energy) states.

This is well described by quantum theory, according to which every stationary orbit conforms to definite value of atom's energy (Bore's postulate). Being placed on stationary orbits, an electron doesn't radiate and doesn't absorb electromagnetic waves. According to the second Bore's postulate, radiation and absorption can take place only when atom changes its state from one stationary state to another:

$$h\varpi_{mn} = h\nu_{mn} = E_n - E_m$$

where ϖ_{mn} or v_{mn} – photon's frequency, E_m, E_n – energy values of states *m* and *n, h* – Planck's constant, *m* and *n* – the numbers of energy states. At the same time, electron jumps from one stationary orbit to another.

Luminescence is defined by the structure of substance energy spectrum, the average time of staying in excited states and rules of selection, which allow absorption or radiation of light of defined frequency. Short-timed luminescence is also called fluorescence. Luminescence which appears during lighting of substance (phosphor) with visible or ultraviolet light is called photoluminescence. Usually process of luminescence satisfies Stokes shift that claims, that wave length λ' of emission light is longer than wave λ of absorption light. According to the quantum theory, this means that photon's energy $h\varpi(hv)$ is used partially for non-optical processes:

$$h\varpi = h\varpi' + E, \varpi > \varpi'$$

where ϖ' – luminescence's frequency, E – energy waste on another process.

Luminescence is characterized by energy output, which equals to ratio of luminescence energy to energy that was absorbed by substance in stable conditions.

Energy efficiency of photoluminescence increases proportionally to wave length λ of absorbed light up to the definite maximum value at $\lambda = \lambda_{max}$ and then rapidly decreases to zero at $\lambda > \lambda_{max}$ (Vavilov's rule). A sharp decrease of energy at $\lambda > \lambda_{max}$ is taking place due to the fact that at these wave lengths λ the energy of absorbed photons is not enough for the process of phosphor atoms and molecules transfer to the excited states.

Ratio of luminescence photons number to absorbed photons with fixed energy is called quantum yield of photoluminescence. According to Vavilov's rule, which is under Stokes shift, quantum yield of photoluminescence doesn't depend on wave length of excited light and rapidly decreases for anti-Stokes radiation.

Intensity of luminescence *I* depends on behaviour of elementary processes, that causes this radiation. In case of spontaneous luminescence, when radiation starts after light absorption, during which atoms or molecules are transmitted to the excited level, that is placed higher than the level at which radiation takes place and then these atoms (molecules) are transmitted to the luminescence level, intensity is subordinate to exponential rule

$$I = I_0 \exp(-t / \tau) \tag{5.1}$$

where I – lighting intensity at the moment t, I_0 – lighting intensity in a moment of excited radiation stopping, $\tau \approx 10^{-9} - 10^{-10}$ – an average duration of excited state of phosphor atoms or molecules. Luminescence of compound molecules and phosphorescence (after lighting) of organic substance are subordinate to the rule (5.1).

Under influence of light there can be happened photochemical transformation of substance (including photosynthesis), which is called photochemical reactions. In a process of such reactions, light absorption takes place. Energy is spent on compound molecules and polyatomic ions decomposition to build parts and on creation of compound molecules of primary ones. An example of photochemical reactions is decomposition carbon dioxide under influence of light

$$2CO_2 + 2h\varpi \rightarrow 2CO + O_2$$

Carbon dioxide decomposition takes place in green parts of plants under sun light influence, as photochemical process, which is a part of photosynthesis.

One of the most important properties of the molecule of chlorophyll, which is the basic pigment of plant cell, is ability to fluoresce. This phenomenon was researched for the first time by Kautsky (Kautsky and Hirsch 1934) and named as chlorophyll fluorescence induction (CFI) or "Kautsky effect".

Dependence of chlorophyll fluorescence induction on time passed after start of lightning of plant's leaves is known as an induction curve or a chlorophyll fluorescence induction curve (Fig. 5.1). The form of this curve is rather sensible to changes in the photosynthetic apparatus of plants during adaptation to different environmental conditions. This fact is a basis for extensive usage of Kautsky effect in photosynthesis research. The advantages of the method of CFI are the following: high self-description, expression, non-invasiveness and high sensibility.

Pigmental complex is the main element of photosynthetic apparatus. Pigments are compound (complex molecules), which absorb light in visible spectral region (400–700 nm). Green pigment of chlorophyll absorbs light in red and blue spectral regions, and in green spectral region this pigment reflects the light. Today there are more than 10 types of chlorophyll known, but the main pigment is chlorophyll "a". Spectral range of chlorophyll "a" is shown in Fig. 5.2.

At the initial moment of time, all channels of photosynthetic transferring of electrons are open and maximum of energy of excited electrons is used for photosynthetic process. At this moment, the chlorophyll fluorescence is minimal and indicated on the Kautsky curve as F_0. Transition $F_I - F_0$ is attributed to reduction of electron-transferring. Transition $F_I - F_0$ is observed, for example, during short period of heat adaptation. Whole stage $F_0 - F_P$ is called as fast stage of fluorescence. Slow stage of CFI includes all inductive transition after peak *P* reaching. Segments of CFI curve are more detailed described in Table 5.1.

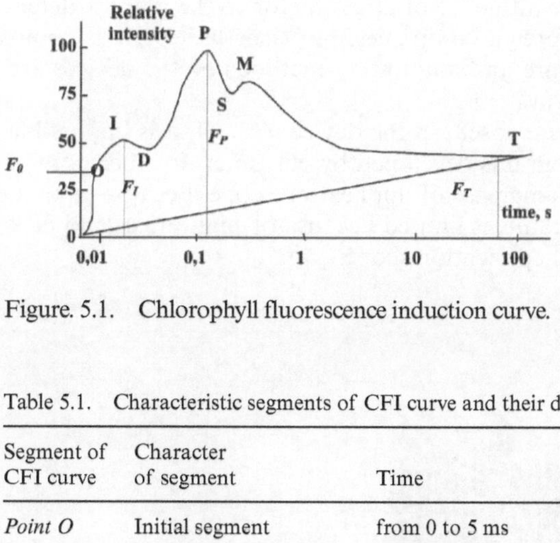

Figure. 5.1. Chlorophyll fluorescence induction curve.

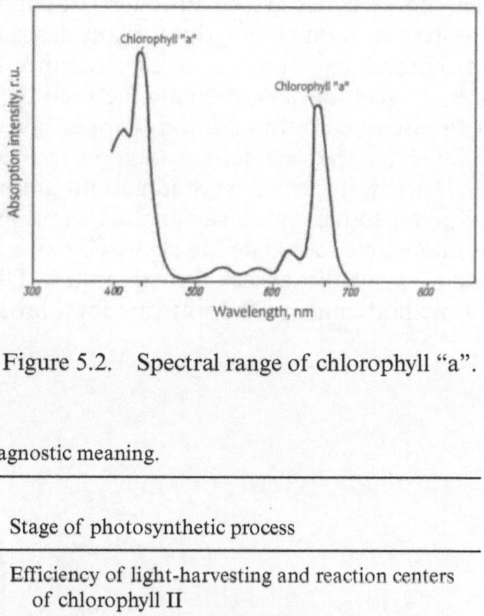

Figure 5.2. Spectral range of chlorophyll "a".

Table 5.1. Characteristic segments of CFI curve and their diagnostic meaning.

Segment of CFI curve	Character of segment	Time	Stage of photosynthetic process
Point O	Initial segment	from 0 to 5 ms (inclusive)	Efficiency of light-harvesting and reaction centers of chlorophyll II
$O-I-D-P$	Attainment of main maximum	0,1 to 10,0 s (0,1 to 1,0 s)	Electron-transferring section (from H_2O to Fd and NADP) – so-called "light stage" of photosynthesis
$P-S-M$	Recession (slope) of curve and attainment of second maximum	3,0 to 50,0 s (0,5 to 10,0 s)	Activation (through Fd) of protein-enzymes of Calvin cycle, setting of pH gradient in membranes, recovery of competing acceptors (O_2, NO_2^- and etc.)
$M-T$	Recession (slope) of curve and attainment of stationary form	10,0 to 300,0 s (10,0 to 2000,0 s)	Mending of reactions of Calvin cycle and flows of matter in plant vessels
$M-S_I$, S_I-M_I-T	Attainment of intermediate minimum	10,0 to 100,0 s	Mending of reactions of Calvin cycle
	Attainment of stationary form	50,0 to 300,0 s (100,0 to 2000,0 s)	Mending of matter transferring through membranes and in plant vessels

Ability to estimate plants states using changing of Kautsky curve form is experimental. So there are examples of changing form of this curve under influence of external factors.

Increase of environment temperature relative to optimal for definite plant type causes decreasing of difference $F_V = F_P - F_0$. It comes as a result of decreasing of activity of electron-transport chain or lighting activity of photosynthesis. During increasing of temperature to destruction level (45–50°C), the level of intensity F_0 increases noticeable. It's possible to choose plants sort that are stable to high temperature influence using these parameters.

Decreasing of environments' temperature relative to optimal for definite plant type causes also decreasing of difference $F_V = F_P - F_0$, because of oppression of photochemical activity photosystem PSII. Ratio $(F_1 - F_0)/F_V$ is increasing. These features allow selecting cold-resistant plants.

Salinity of ground results in decreasing of level of F_P and F_0. Ratio $(F_P - F_0)/F_P$ is decreasing. It comes as a result of oppressing of photo system PSII activity. Using these features it's possible to choose plants that are stable ground salinity.

Water deficit results in decreasing of subtraction $F_P - F_0$ in direct proportion with decreasing of water potential of leaf. Most probable reason is slowing down of photo system PSII recovery of primary acceptor because of oppression of excretion oxygen and intersystem transport of electrons.

5.3 PORTABLE DEVICE "FLORATEST"

The portable device of "Floratest" family (see Fig. 5.3) measures intensity of chlorophyll fluorescent induction without plant destruction. By the form of curve of chlorophyll fluorescent induction (similar to cardiogram) the influence of stress factor to the plant is determined. Such devices are manufactured by foreign companies, but their unit price is about 1 thousand euro. In addition, applied software and amount of methods of the devices are limited and they are designed as constrained systems.

Thereby, if needed method and program are absent in the device manual, it is impossible to add it to the device new methods and programs developed by customer. Furthermore, it is impossible to update the devices by user to increase of application scope. Because of that, the usage of the devices, for example, in Ukraine, is limited because of high prices and lack of methods and applied software for a broad application scope.

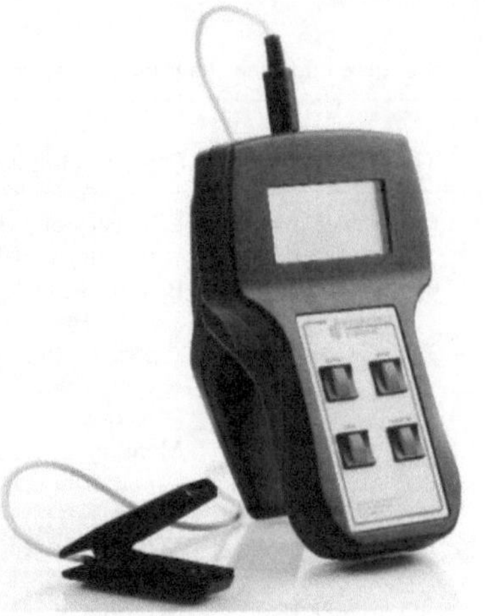

Figure 5.3. Portable device of "Floratest" family.

The firms in USA, Canada, Western European countries manufacture portable devices for measuring chlorophyll fluorescent induction to diagnose of a plant state. There are such well-known companies among them as Opti Sciences, USA, (www.optisci.com), Hansatech Instruments, Great Britain (www.hansatech-instruments.com), Heinz Walz, Germany, (www.walz.com), EARS, Netherlands, (www.ears.nl).

The competitive performances of developing portable device "Floratest" and competitors are given in the Table 5.2.

Photosynthesis and induction of fluorescence of chlorophyll operate in a direct competition for a finite pool of absorbed light energy. Any change in energy utilization by one process produces a complementary change in the others. So this fact enables induction of fluorescence of chlorophyll to be used as a prompt and reliable non-invasive estimation of photosynthesis.

Fluorometers are intended specifically to detect the chlorophyll fluorescence emission from a leaf of plant. The main features of new design are wide dynamic range, high precision and speed, low quiescent current and low operating voltage. The device is to be designed as an open system, user reprogrammable and easy-to-use. To be used in field conditions, the new design has to be portable and be equipped with a battery with low operating voltage. "Floratest" hardware architecture is shown in the Fig. 5.4.

Table 5.2. The competitive performances of developing biosensor device and competitors.

Device name	Floratest	OS-30p	CL-01	HANDY-PEA	PAM-2 100	PPM-100
Company	V.M. Glushkov Institute of Cybernetics of NAS of Ukraine	OPTI-Sciences	Hansatech Instruments	Hansatech Instruments	Heinz Walz GmbH	EARS
Type of device	Portable	Portable	Portable	Portable	Portable	Portable
Wavelength of irradiation, nm	670	660	620 940	660	655	637
Measuring time, s	10…240	2…255	0.5…60	0.1…300	0.1…300	0.1…300
Ability of reprogramming and changing of applied method	yes	no	no	no	no	no
Price, euro	275	2495	1275	1350	1500	1395

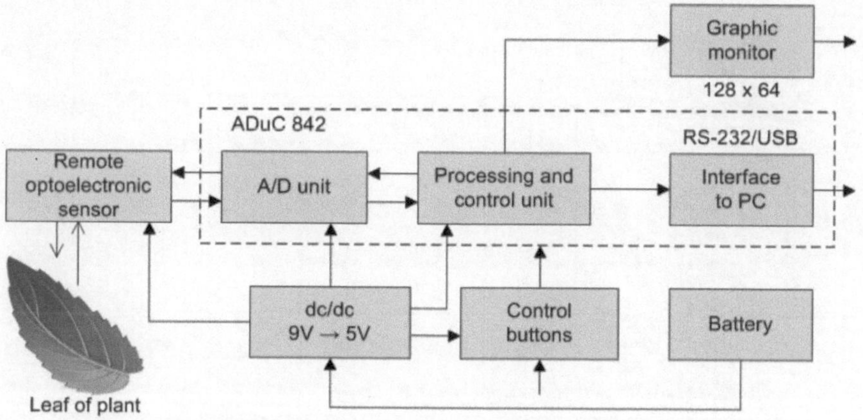

Figure 5.4. Hardware architecture of developed portable device "Floratest".

According to Fig. 5.4 device "Floratest" includes high-integrated microprocessor unit and small remote optical sensor. The main features of the remote optical sensor: light and low-powered. Simplified functional diagram of remote optical sensor is shown in Fig. 5.5.

The optical sensor consists of blue light-emitting diodes and a photodiode amplifier. The type of diodes is FYLS-0603UBC (http://www.foryard.com). The sizes of diode are 0.6×0.3 mm. Diodes are compatible with automatic placement equipment and reflow solder process. Luminous intensity of the diode is 120 mcd, viewing angle 130 degree and dominant wavelength 470 nm. Typical electrical characteristics curves of blue diode FYLS-0603UBC are shown in Fig. 5.6, Fig. 5.7 and Fig. 5.8.

The remote optical sensor is built as a "reflection diagram" on the basis of four light-emitting diodes and one photodetector. "Reflection diagram" means that light-emitting diodes and photodetector are situated on the same side of researched leaf. To research chlorophyll fluorescence in the red spectral region, the filter is placed on the input of photodetector. Emission intensity of light-emitting diodes and photodetector sensitivity can be changed during a measuring process.

The type of the photodiode amplifier is TSL251R (http://www.taosinc.com), combining a photodiode and a transimpedance amplifier on a single monolithic integrating circuit. These devices have improved amplifier offset-voltage stability and lowered power consumption and are supplied in a 3-lead clear plastic package with an integral lens. Functional block diagram of TSL251R is shown in Fig. 5.9. The device is a single supply voltage operation with

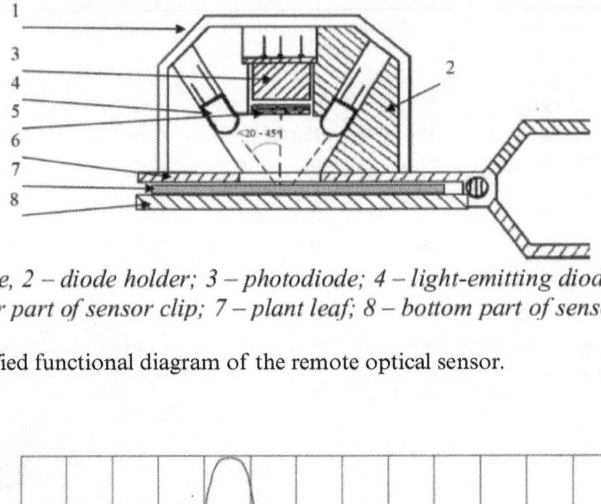

1 – sensor enclosure, 2 – diode holder; 3 – photodiode; 4 – light-emitting diode; 5 – optical filter; 6 – upper part of sensor clip; 7 – plant leaf; 8 – bottom part of sensor clip

Figure 5.5. Simplified functional diagram of the remote optical sensor.

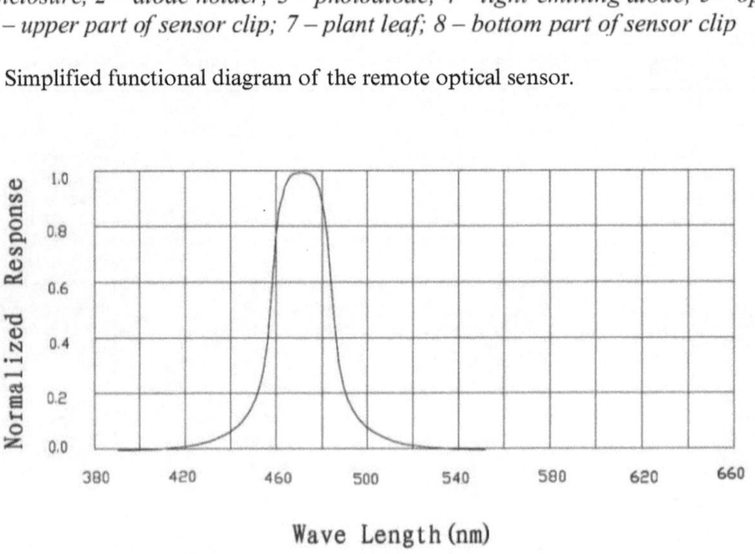

Figure 5.6. Spectral characteristic of blue diode FYLS-0603UBC.

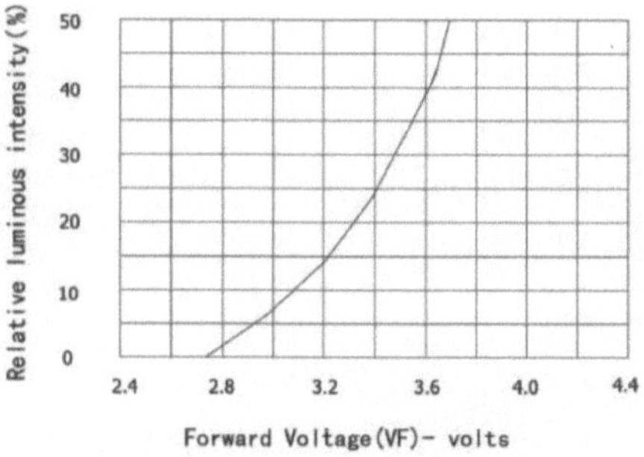

Figure 5.7. Relative luminous intensity vs Forward voltage of blue diode FYLS-0603UBC.

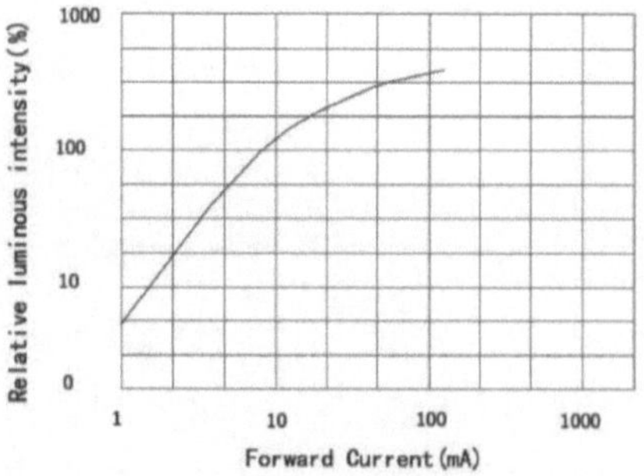

Figure 5.8. Relative luminous intensity vs forward current of blue diode FYLS-0603UBC.

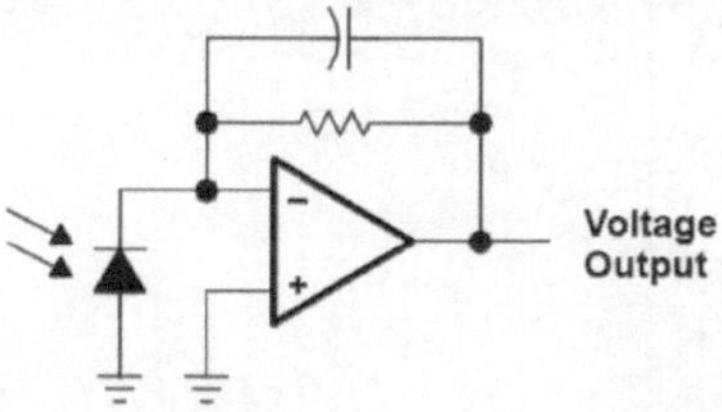

Figure 5.9. Functional block diagram of photodiode TSL251R.

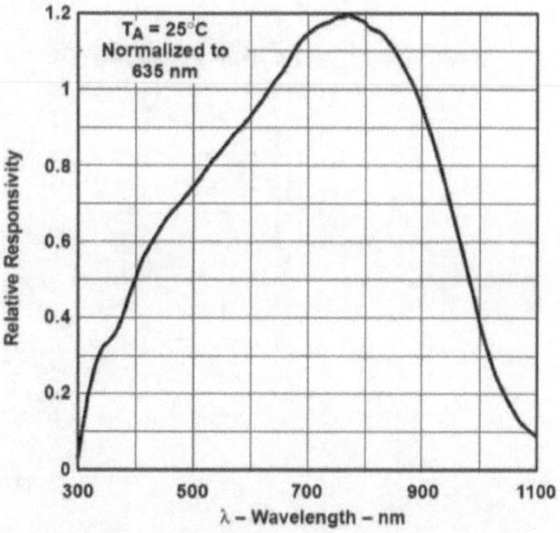

Figure 5.10. Spectral responsibility of Photodiode TSL251R.

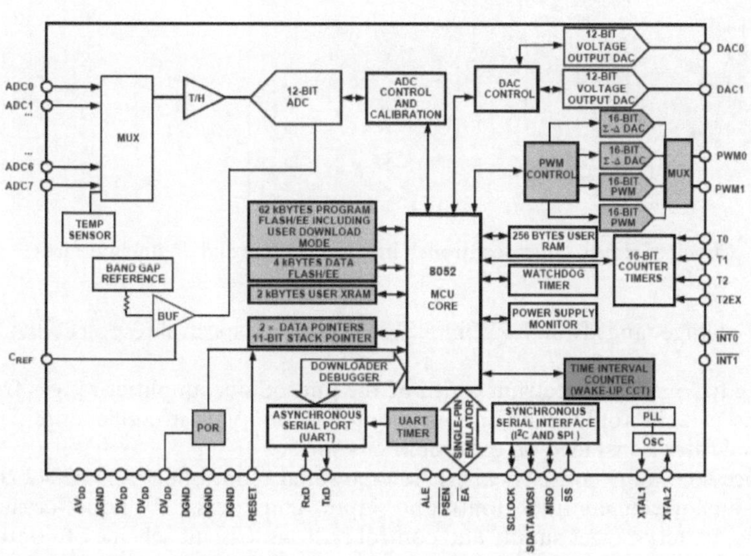

Figure 5.11. Functional block diagram of microcontroller ADuC842.

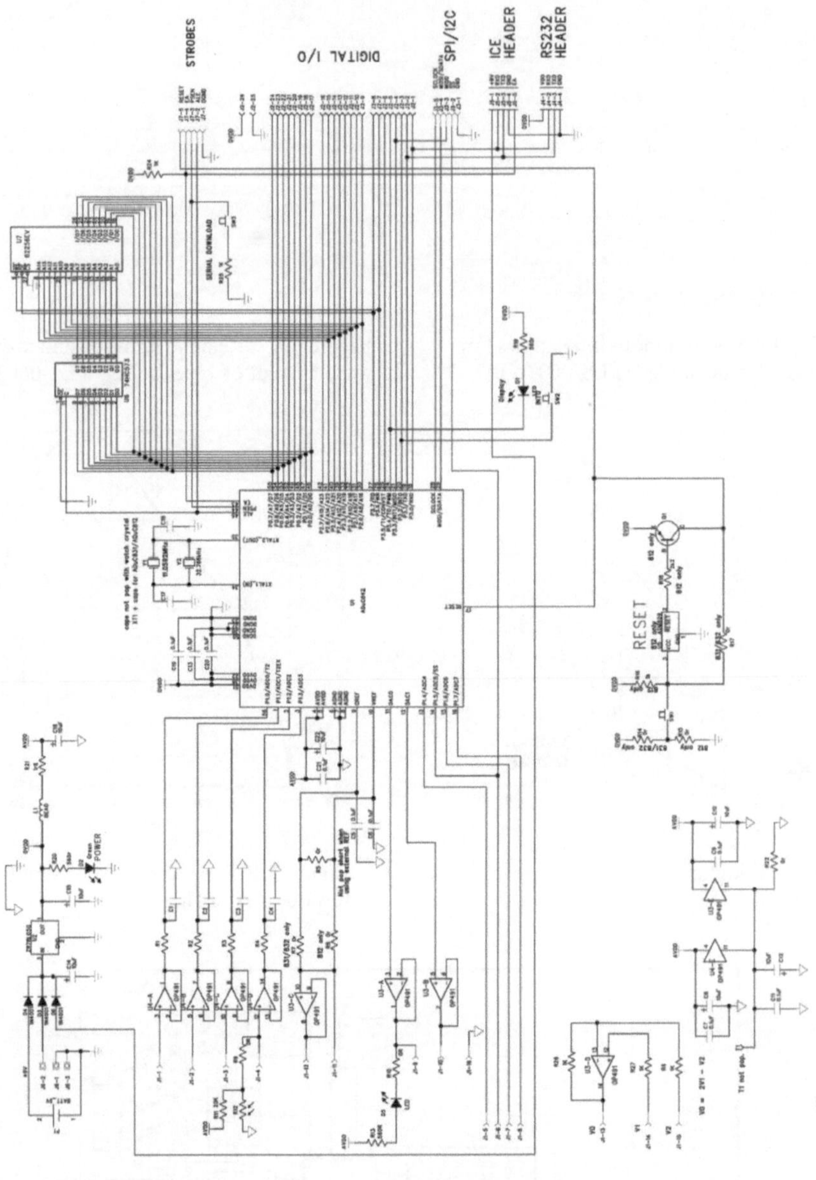

Figure 5.12. Circuit scheme of microcontroller, input amplifiers and digital interfaces.

wide supply-voltage range from 2,7V till 5,5V. Photodiode spectral responsibility is shown in Fig. 5.10.

According to Fig. 5.10, spectrum width of the photodiode amplifier ranges from 300 nm to 1100 nm. Peak chlorophyll emission is about 700 nm. A photodiode amplifier spectrum narrowing red filter is installed on the amplifier's surface.

Microprocessor unit consists of a low-powered controller ADuC842 (Fig. 5.11) (http://www.analog.com/en/index.html) and input amplifiers, 3V liquid-crystal monitor FDCG12864, battery-power supply and control buttons. Circuit scheme of controller, input amplifiers and digital interfaces are shown in Fig. 5.12. Printed circuits boards of microprocessor unit are shown in Fig. 5.13 and Fig. 5.14.

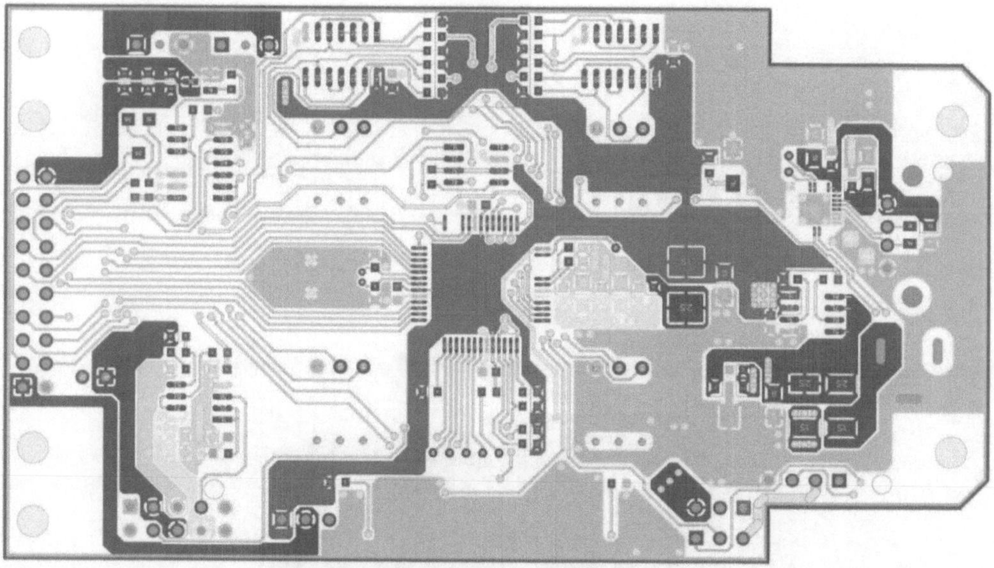

Figure 5.13. Printed circuit board of microprocessor unit.

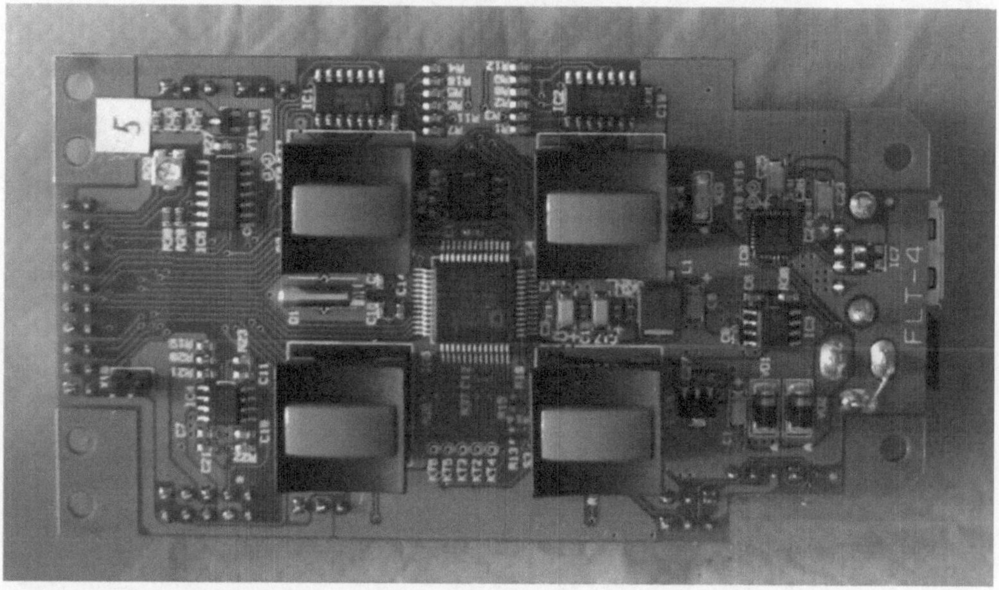

Figure 5.14. Real printed circuit board of microprocessor unit.

The real benefit of the microprocessor unit is its high performance, low power consumption and small-scale size. All electronic components are selected from electronic libraries of virtual laboratory of automation design (Palagin et al. 2010) developed in the V.M. Glushkov Institute of cybernetics of NAS of Ukraine. They are intended for surface mounted assembly technology, 3V operating voltage, designed in small packages.

Laser prototype technology was used to design and manufacture remote sensor enclosure. This technology can offer the competitive edge in getting products to the market faster from

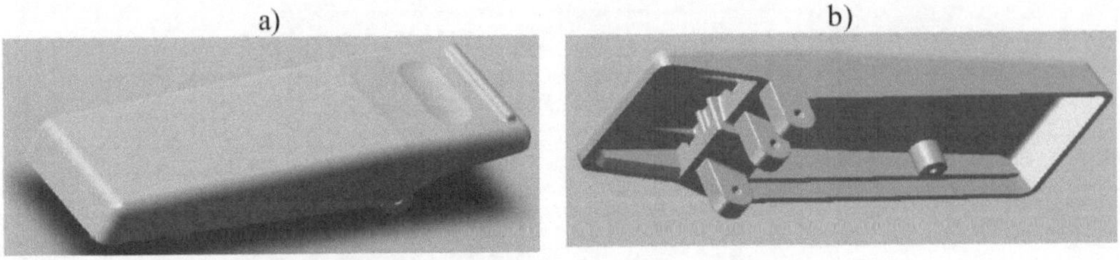

Figure 5.15. 3D-model of remote sensor enclosure: a – top, b – bottom.

Figure 5.16. Manufactured remote sensor enclosure.

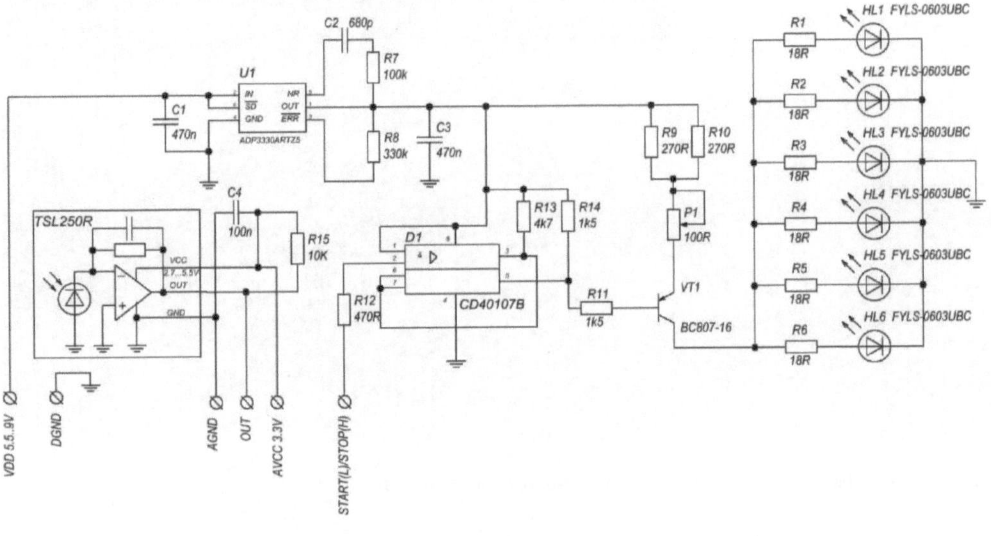

Figure. 5.17. Principal electronic scheme of remote optical sensor.

early design and concept models to skillfully finished engineering models and low volume production runs. 3D-model of remote sensor enclosure is shown in Fig. 5.15. Manufactured remote sensor enclosure is shown in Fig. 5.16. Now we use technology of 3D printing by ABS plastic to produce prototypes or small batches of sensor enclosure.

Remote sensor consists of two printed circuit boards and red filter installed on the optical window. Principal electronic scheme of remote optical sensor is shown in Fig. 5.17. Two designed printed circuit boards of remote sensor are shown in Fig. 5.18. Two manufactured printed circuit boards of remote sensor are shown Fig. 5.19. Relative position of two printed circuit boards of remote optical sensor is shown in Fig. 5.20.

Supporting of power supply of portable devices is a very important task. Therefore, for more effective work of device, small printed circuit board for microprocessor unit was

Figure 5.18. Designed printed circuit boards of remote sensor: a) – top printed circuit board, b) – main printed circuit board.

Figure 5.19. Manufactured printed circuit boards of remote sensor: a – top printed circuit board, b – main printed circuit board.

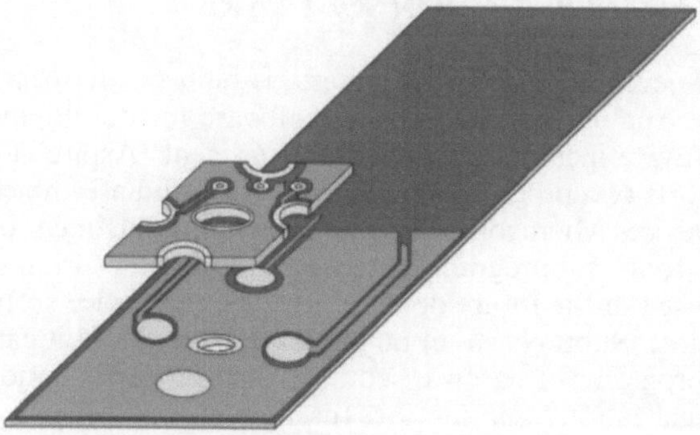

Figure 5.20. Relative position of two printed circuit boards of remote optical sensor is shown.

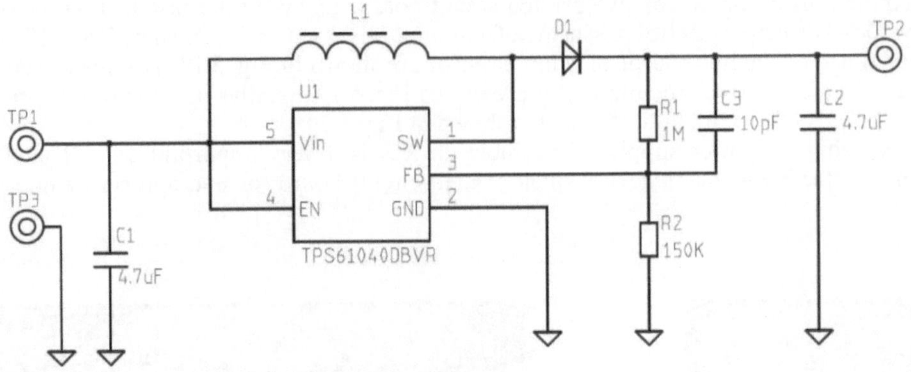

Figure 5.21. Principle electrical scheme of dc/dc-converter.

Figure 5.22. Printed circuit board of dc/dc-converter.

developed and performed, which converts 5V DC into 9V DC. Now this printed circuit board is embedded in the device enclosure as a separate component, and in the future this board will be built in main printed circuit board of microprocessor unit. Principle electrical scheme of the dc/dc-converter is shown in Fig. 5.21 and manufactured printed circuit board of dc/dc-converter is shown in Fig. 5.22.

5.4 CREATING MAIN APPLICATION SOFTWARE

As mentioned above, developed device "Floratest" is built on the basis of microcontroller ADuC842. Company Analog Devices provides software for development and adjustment of this hardware. Software includes integrated environment "Aspire" for developing device software, which supports two programming languages (C and assembler), and environment "ADSIM" (Analog Devices Microconverter Simulator) for adjustment of device software.

By means of these tools, subprogram library for fluorimeter "Floratest" was developed. This library may be used in the future developments of the device software for diagnostics of biological object state. Subprogram library is an open tool and it can be complemented. Library includes subprograms of hardware control, data processing and exchange, measurement of the result visualization, user interface (user menu).

Structure of "Floratest" software is shown in Fig. 5.23. "Floratest" software includes next components: programs for user interface supporting, device operation tests, device

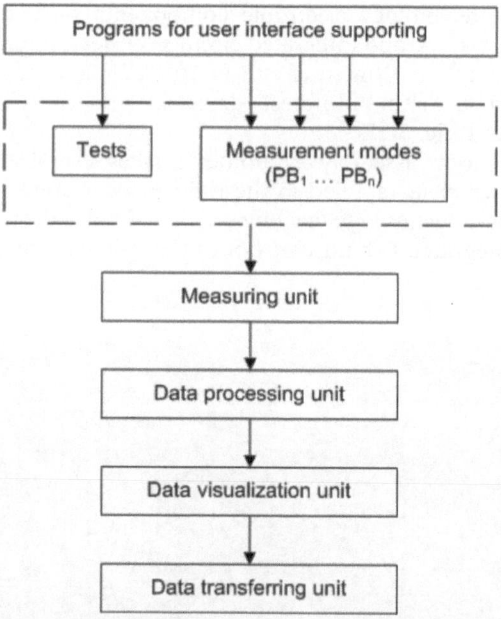

Figure 5.23. Structure of "Floratest" software.

measurement modes ($PB_1...PB_n$), measuring unit, data processing unit, data visualization unit, data transferring unit.

Programs for user interface include programs for window menu displaying on liquid crystal display and programs for item menu choosing. Programs for user interface supporting form data structure, which defines selected measurement mode $PB_1...PB_n$, in the memory. This data structure consists of the following data: measurement number, measurement duration and buffer for measurement data. In the data structure, there are also reserved sells for values of characteristic quantities, which are defined during measurement.

A measuring unit includes programs to control of ADC operation and measurement process and programs for synchronization of stimulating effect on a biological object. In the measuring unit the measured background component is defined and then subtracted from measured values.

In data processing unit the values of characteristic quantities are defined and necessary proportions are calculated: F_0/F_m; $(F_m - F_0)/F_m$; $(F_m - F_{st})/F_m$.

Programs of data visualization unit control outputting of results on the LCD. They consist of fonts and programs to clear of display, visualization of obtained result in the graph mode, selection of graph zoom in accordance with signal maximum value and calculation of graph point coordinates.

Data transferring unit provides adjustment of serial port on certain data receiving and transmission speed from and to PC.

5.5 CONTRACT MANUFACTURE

It was used electronic contract manufacture of VD MAIS firm (Kyiv, Ukraine) (http://www.vdmais.kiev.ua) for creating serial party of the portable biosensor of "Floratest" family.

So the main features of technical documentation were matching functional diagrams of device units with interoperability of Surface mount technology assembly line. There were selected Surface mount technology electronic components such as resistors, capacitors, inductances, diodes, transistors, and integrated circuits for device units and on this basis the technical documentation of the Printed circuit boards of microprocessor unit and remote optical sensor was designed. Serial "Floratest" device kit and CD with software and drivers are shown in Fig. 5.24 and Fig. 5.25.

According to Surface mount assembly technology, it takes several steps for the bare Printed circuit boards with solder paste, applied in the right places before becoming fully-fledged boards. The Surface mount technology machines pick and place the tiny resistors, solid-state capacitors and other Integrated Circuit chips onto the Printed circuit boards at ultra-high speeds.

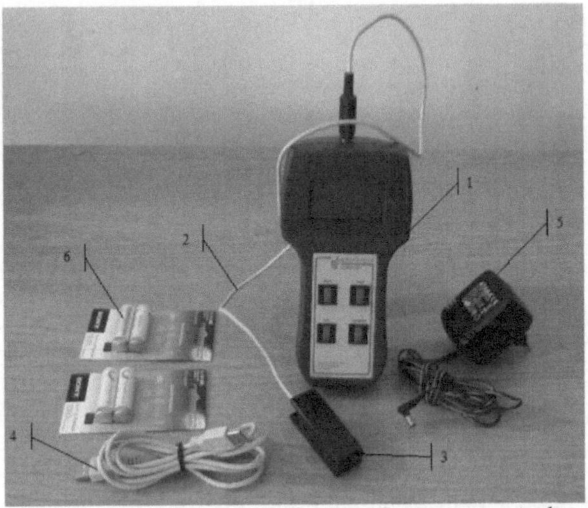

1 – microprocessor unit, 2 – cable of remote optical sensor, 3 – remote optical sensor, 4 – data cable, 5 – power supply unit, 6 – four rechargeable batteries

Figure 5.24. Device "Floratest" kit.

Figure 5.25. The appearance of CD with software, drivers and user documentation.

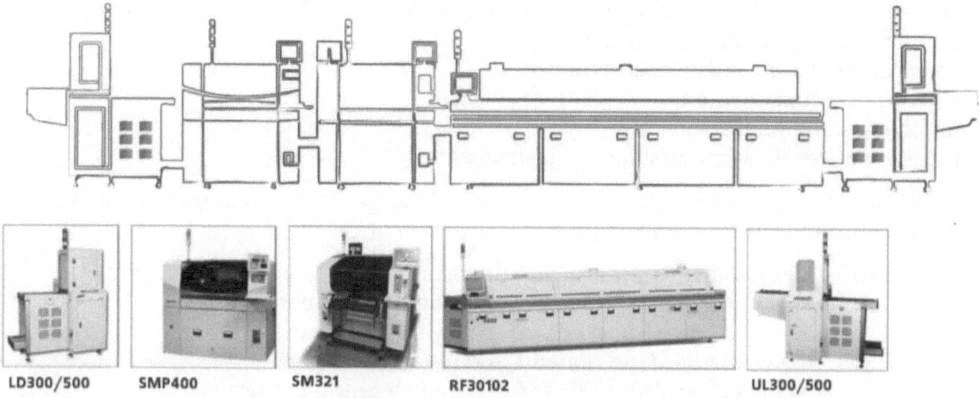

1 2 3 4 5

1 – automatic loader of boards, 2 – stencil printer, 3 – automatic placer, 4 – conventional-conveyor oven, 5 – automatic unloader of boards

Figure 5.26. Structure and composition of contract manufacture technological line.

If you look for example at the motherboard in computer, some of these small components are no more than 1 mm square. Each board passes through two sets of Surface mount technology high-speed machines, the 'small pick and place' and 'large pick and place' devices. Each machine in the set adds a few components, from tiny resistors up to Integrated circuit chips. Using belt fed from tape-like cartridges of components, the Surface Mount Technology gear places components like a machine gun, needing as little as an eighth of a second to place a component with exact precision on the Printed circuit boards.

During the manufacturing of printed circuit boards of all units of the device, there was fine-tuned assembly technological processes and technical documentation too. For the perfection of documentation and manufacture technology, there were created more than one hundred printed circuit boards of all types.

Structure of contract manufacture technological line and its composition are shown in Fig. 5.26.

5.6 OPTICAL SENSOR TESTING

Leaf structure and chlorophyll structure ensure certain optical properties of native chlorophyll, which have to be considered during working out testing methods for sensor of the "Floratest" device.

Testing of device sensor is necessary to compare measuring results and to determine values of different devices. Nowadays, the sensor testing and calibration is made during device manufacturing, but sometimes it can be executed during the device operation.

The most suitable method among known testing methods of measuring is using standard instruments and simulators of measuring value, notably singular standards of fluorescence, absorption and reflection of native chlorophyll.

Using of standard instruments for sensor testing is complicated by impossibility of ensuring equal metrological and technical conditions for standard instruments and tested sensors. So, for sensors of portable devices (like "Floratest" device), there are such features:

1. different sizes of illuminated spot (0.5–10 mm);
2. necessity of native chlorophyll and leaf intactness;
3. heterogeneity of optical properties within bounds of one leaf;

4. variability of functional characteristics, namely chlorophyll fluorescence induction;
5. difference of the result presentation;
6. constructive disagreement with sensor unit;
7. absence of testing methods;
8. multitype standard instruments;
9. necessity of certification of standard instruments.

For sensor testing, it is reasonable to use standards of fluorescence, absorption and reflection, which meet the next requirements:

1. homogeneity and stability of optical characteristics within bounds of the standard;
2. independence of optical characteristics from environmental conditions;
3. small size, weight and constructive compliance with sensor unit;
4. possibility to use in standard instruments and tested devices;
5. single-type of standard optical characteristics and measuring value;
6. possibility of calibration without using of plant leaf;
7. possibility and convenience of testing in field conditions;
8. possibility of standard certification;
9. low cost of standard.

Standards for testing of sensors of developed devices have to simulate optical characteristics of chlorophyll and consider technical features of tools for measuring of these characteristics (see Fig. 5.27).

In developed "Floratest" device, the photoreceiver with filters has to be tested. Standard for such testing is a source of light with spectral characteristics, which are similar to chlorophyll optical characteristics. Standard is placed in sensor on the place of plant leaf. Spectral characteristics of testing can also be provided by three passive standards, which have optical characteristics similar to optical characteristics of chlorophyll.

The requirements to chlorophyll fluorescence standard mentioned above can be met by glass plate ($15 \times 15 \times 2$ mm), placed in the sensor instead of plant leaf. When standard is illuminated by light intensity of 50 W/m^2, the device has to show straight line parallel to time axis, equal 0.6 (quantum yield) of maximum measurement range. This value can be reached by changing amplifier gain of photoreceiver. Such standard using for testing of sensor of the developed "Floratest" device lets to equalize amplitude characteristics during manufacturing of pilot party of devices.

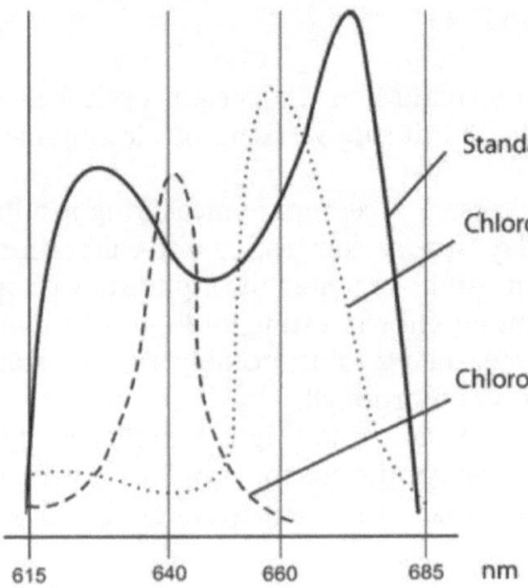

Figure 5.27. Spectrums of absorption.

Normative values of chlorophyll concentration for optical absorption by standard on proper wavelengths are determined by parallel photocolorimetric determination (Bulatov and Kalinkin 1972) of chlorophyll extracts and standard of chlorophyll absorption.

Also during testing of reflection type sensor, there are used universal integral standards of reflection. Determination of reflection coefficients of leaf is fulfilled in wavelength range from 310 to 1100 nm. System UniCpec-SC proposes some white standards of maximal reflection.

As standards of reflection (WWW/PPSYstems.cjm), plates covered by paint are used. Maximal reflection is given by paint on the base of magnesia, and minimal—on the base of soot.

Testing includes next operations:

1. Checking device functionality after switching on according to operating manual.
2. Setting necessary modes.
3. Execution of one measurement without standard. Device sensor gives black current.
4. Standard placing and execution of one measurement:
   - chlorophyll sensor show straight line, parallel to X axis and equals 0.65 of maximum scale;
   - absorption type sensor gives values of passing red and infrared light, and the difference with incident signal gives absorption signal;
   - reflection type sensor gives signal of maximum reflection on selected wavelength and signal of minimum reflection.
5. Measurement of the results on standards compare with normative values for these standards and estimate necessity and value of correction.

5.7 APPLICATION OF THE "FLORATEST" DEVICE IN AGRICULTURE AND ECOLOGICAL MONITORING

Portable device "Floratest" is used for real-time monitoring of plant state. The device operation is based on the chlorophyll fluorescent induction method mentioned above, and provides the estimation of photosynthesis process quality, which is the main process of plant vital activity. The "Floratest" devices are now used as a part of data acquisition systems or independently for working out the methods for solution of following tasks:

1. prompt estimating of plant life activity after drought, freeze, imps, pesticide application, impacts of climate fluctuations and stress factors of natural and anthropogenic origin etc.;
2. prompt estimating of optimal dozes of fertilizers, biological active admixtures and humic substances. It makes possible to increase protective attributes of plants and trees to the climate fluctuations, influence of stress factors of natural and anthropogenic origin and reduce the content of harmful substances in soil and plants;
3. express estimating by indirect methods of soil, water and air pollutions by pesticides, heavy metals, industrial emissions and their impact on green plantations in megalopolis;
4. ecological monitoring of state of megalopolis green plantations;
5. profound scientific research in area of ecological monitoring and environmental protection in universities and schools.

The experimental researches of the "Floratest" were conducted in National Scientific Center "V.E.Tairov's Institute of viticulture and winemaking" of National Academy of Agrarian Sciences of Ukraine (Romanov 2007) and several farming in Ukraine and Germany.

Examples of the practical usage of fluorometer "Floratest" in the National Scientific Center "V.E.Tairov's Institute of viticulture and winemaking" are shown in Fig. 5.28 and the graph of CFI on the device display are shown in Fig. 5.29.

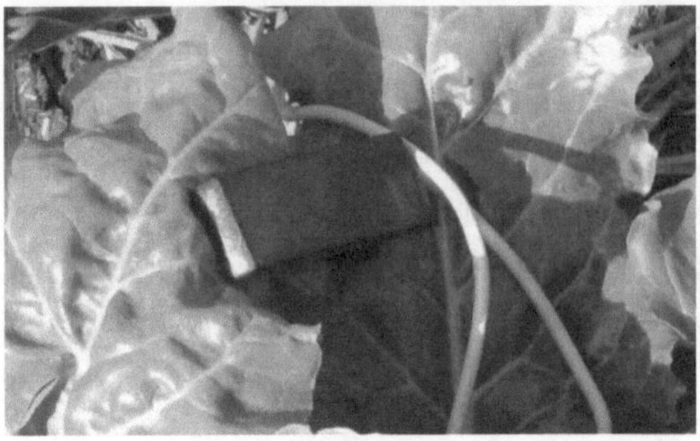

Figure 5.28. The sensor of the "Floratest" on the leaf.

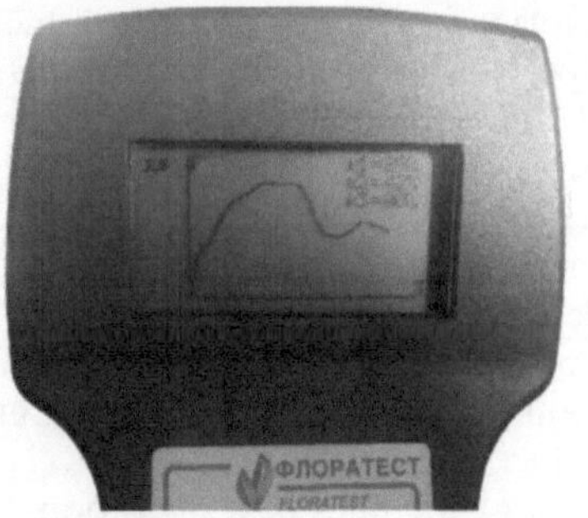

Figure 5.29. The image of CFI on the device's display.

The conditions and results of the experimental researches are listed below.

Mature leaves of vine were used in the researches. Under changes of soil watering conditions, there were observed sharp changes in behavior of induction transitions of chlorophyll fluorescence which were accompanied by quite essential changes of leaf tissue spectral characteristics.

Determination of fluorescence spectral characteristics was done by placing the device's sensor on the leaf's surface without integrity disturbance directly in a pot or in a field. It allowed to research on plastid and vacuolar pigments in their natural state and in that way approaching to understanding of the biophysical and physiology-biochemical processes which take place in the live leaf, and determination of important sides of photosynthetic activity.

Fluorescence intensity of the sample was determined in relative units.

It is significant that under natural conditions in the middle latitudes the drought is accompanied simultaneously by high temperatures of air, and that intensifies bad influence of ground water lack on agricultural plants.

Even in the first variant of experiment (drought), considerable changes of the behavior of fluorescence induction comparing to the control samples appeared. Changes appear in weakening of penetrability of the chloroplasts' membrane structures. That results in substantial increase of time characteristics of fluorescence induction slow decrease. At the same time, noticeable variety differences become apparent. Sharp decrease of its value is typical for profound functional injuries of photosynthetic structures and cells of particular variety entirely.

Accordingly, in this stage of drought influence significant variety differences in exsiccate factor resistance of both photosynthetic structures and lamina's parenchymal cells entirely became apparent.

More deep changes of destructive nature may be observed in case of high temperatures (+40°C), which have an influence on leaves complementary to drought. In this case for all the varieties being studied significant and almost irreversible functional changes of plastid structures are noted. These functional changes show in sharp decrease of CFI intensity.

Disastrous changes of life activity of vine leaf cells which take place during these processes show in oppression of biosynthetic processes, intensive decomposition of cytoplasmic structures and intensification of oxide catabolism of plant cell's content. The consequence of these processes is decrease of CFI intensity as a result of its oxidizing transformation.

In the Fig. 5.30 there are shown the diagrams of measuring of chlorophyll fluorescence intensity for vine plants (sort Kober 5BB) during 5 months. The vine plants were under drought influence and normal conditions.

Thus, a water deficit shows up on the Kautsky curve as difference of fluorescence (*FP—F0*) decrease. The most credible reason of this is oppression of oxygen emission which is related with slowing down of electrons transfer. Assuming that *F0* almost does not change for the

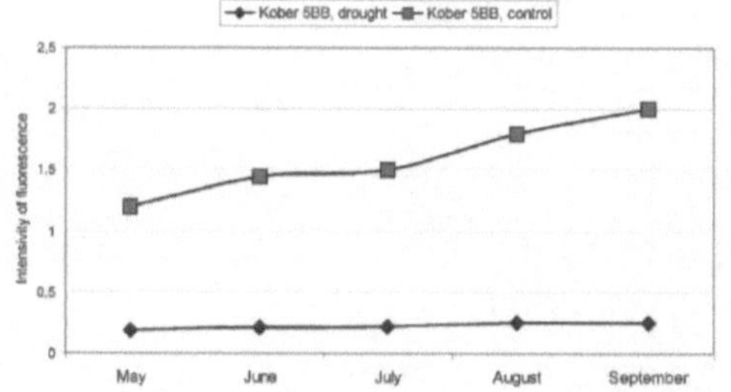

Figure 5.30. CFI intensity of vine plant (Kober 5BB) under drought and normal conditions.

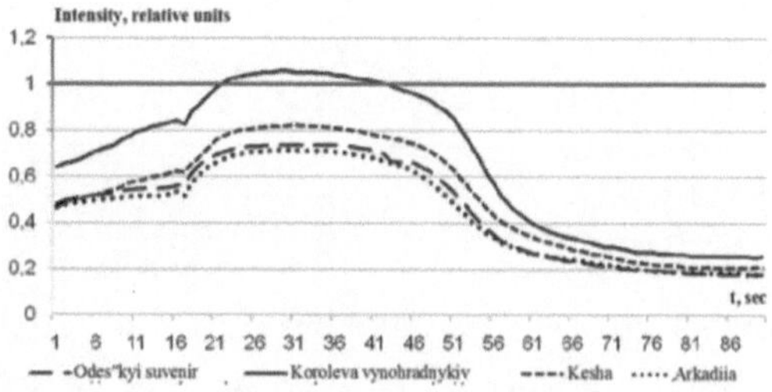

Figure 5.31. Measuring of chlorophyll fluorescence induction curve before predrying.

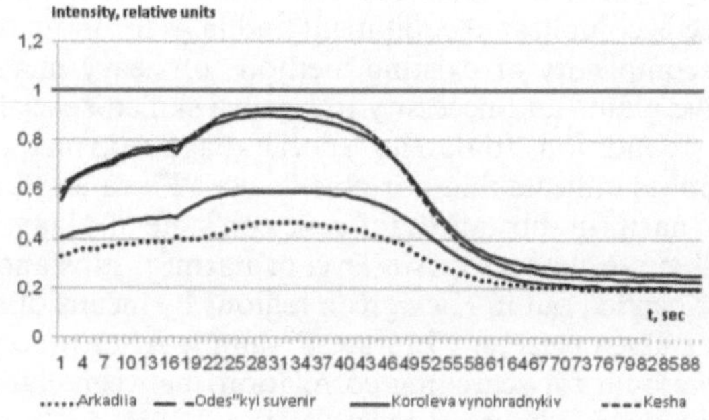

Figure 5.32. Measuring of chlorophyll fluorescence induction curve after 10 hours of predrying.

Figure 5.33. Measuring of chlorophyll fluorescence induction curve after 24 hours of predrying.

test and control plants, in a maximum point the chlorophyll fluorescence intensity value can define the level of water deficit.

In Figs. 5.31–5.33 there are some results of researches for detecting of water deficit of plants, which were executed by means of portable device "Floratest" in the National scientific centre "Institute of viticulture and wine-making named after V.Ye. Tairov" of National academy of agrarian sciences of Ukraine.

Together with Megalopolis Ecomonitoring and Biodiversity Research Center of National Academy of Sciences of Ukraine, experimental researches of portable device "Floratest" were carried out to detect influence of heavy metals and harmful substances (e.g. lead, sodium, chlorine etc.) in leaf and soil on the plants state in Kiev. Today long-term phytomonitoring

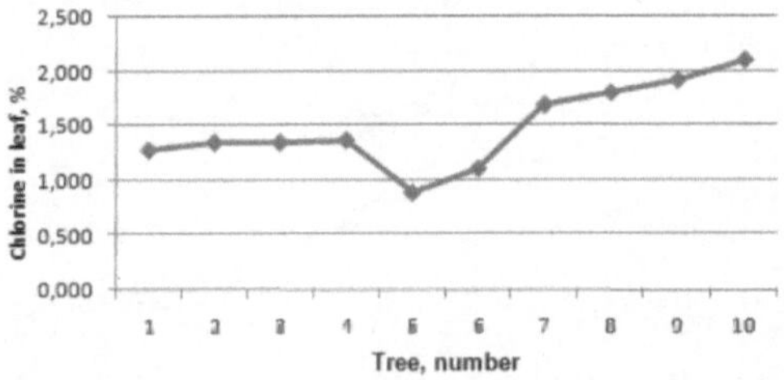

Figure 5.34. Chlorine content in researched trees' leaf.

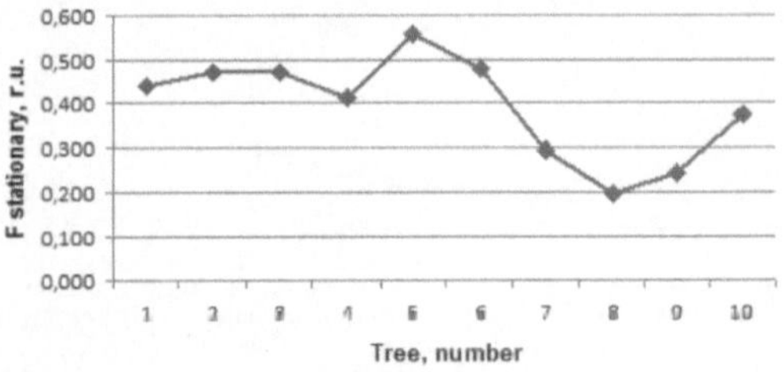

Figure 5.35. Readouts of portable device "Floratest" for researched trees.

methods are used. They consist of visual observations and chemical analysis of soils and plant fragments and needs complicated equipment and lasts for more than one week.

Long duration and complexity of existing methods of heavy metals and harmful substances detecting in alive plants and necessity to involve skilled specialists to perform these researches set necessary conditions to develop special diagnostic methods and tools for this aim. Preliminary researches indicate that portable device "Floratest" can be used for detecting of heavy metals and harmful substances influence on state of plants by measuring of CFI curve. The form of CFI curve changes versus level of harmful substances influence.

Joint researches were carried out in Kiev green regions by means of common phytomonitoring methods and portable device of "Floratest" family. After processing of the research result, the relations were built for searching correlations between chlorine content in trees' leaf, which are got by common phytomonitoring methods (Fig. 5.34), and readouts of portable device of "Floratest" family (Fig. 5.35).

Even one can see some dependence between chlorine content in trees' leaf and readouts of portable device "Floratest" (stationary segment of CFI curve) in Fig. 5.34 and Fig. 5.35. Calculations, made by mathematical methods, show certain correlations between these values. Such researches were made for other harmful substances, such as sodium, magnesium.

It is easily to concede that CFI curve form expresses not only contents of separate harmful substance, but general state of tree versus influence level of harmful substances.

5.8 DISTRIBUTED DATA ACQUISITION SYSTEM

Development of electronics and telecommunication technologies causes creating and implementation of electronic data acquisition systems with remote gathering of measured data in many areas of human activity. Such systems include smart sensors, communication lines, transmitters and receivers, central control stations, which gather and accumulate information.

Nowadays, it is impossible to imagine the precision farming technology without such an important element as decision-making and agrotechnological operation generating system. Data from system of primary sensors, placed directly on agricultural fields, can be used as a basis for decision making. Quality and completeness of data acquisition determine the correctness of decision making about implementation of necessary agrotechnological operations for harvest increasing and costs reducing.

It is necessary to note that functional possibilities of decision-making system in precision farming are determined significantly by quantity, quality and frequency of data acquisition from fields and data processing. In general form the upper level of decision-making system in precision farming is shown in Fig. 5.36.

The union of mobile and traditional data acquisition technologies, database of agricultural fields and software for knowledge formalization and knowledge presentation can be

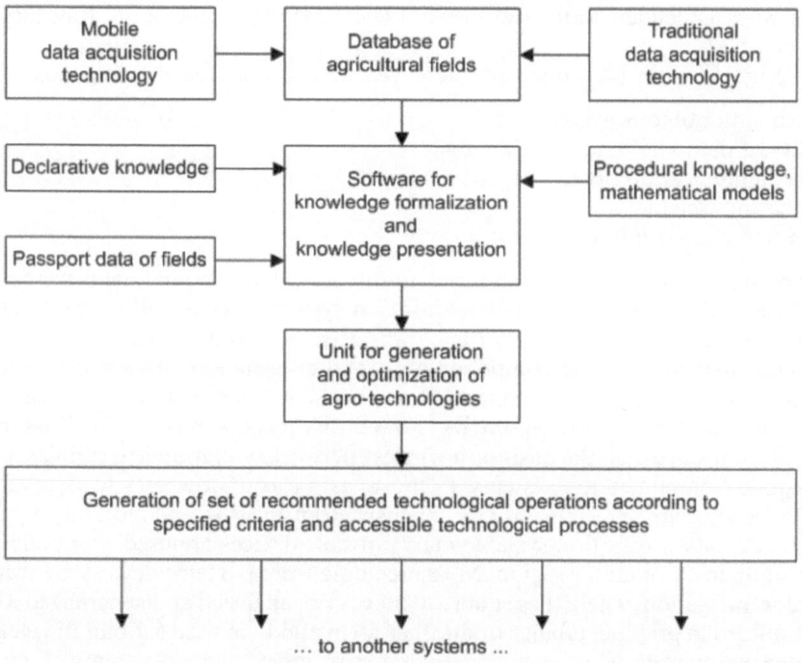

Figure 5.36. Upper level of decision-making system in precision farming.

named as the base agrotechnology. So the next part of this system, namely unit for generation and optimization of agrotechnologies, can be named as the adaptive agrotechnology.

Meteorological data, information about soil state and, not less importantly, information about state of plants, can be referred to data which are acquired directly on the field.

Therefore, acquisition of live and objective data of plant cover state in most cases is very important factor, which causes future strategy of keeping agricultural lands and proper decision making. Certainly, it would be ideal to obtain information about improvement or worsening of plant cover state beforehand, but not after the event. It lets to avoid increasing costs and save harvest from possible loss.

As it is shown in Fig. 5.36, the mobile and traditional data acquisition technologies play an important role in precise farming as primary data sources. For this aim, wireless data acquisition system which obtains information about plant state on large territories of agricultural fields, woodlands or parklands in real-time (Palagin et al. 2011a) is developed in V.M. Glushkov Institute of Cybernetics of NAS of Ukraine.

As a sensor to obtain information about a plant state, in such system we use tiny wireless sensors and portable wireless devices of "Floratest" family (Palagin et al. 2011b), which were developed by us. We used portable device of "Floratest" family, which is shown in Fig. 5.3, as the first version of data acquisition unit.

Field testing devices of "Floratest" family, including in agrarian farming of Germany, showed some restrictions of its application. The data reading from one plant lasted for some minutes. But monocultures are growing in large agricultural areas and due to get data from these areas for crop monitoring, it is necessary to make thousands of measurements. For this purposes, there are dozens of the devices and large service staff of high-level skill needed.

To develop technology of prompt estimation of plant state based on CFI method for precision farming it was designed and prepared for serial production remote smart wireless biosensors with embedded radio transceiver. The functional scheme of tiny biosensor is shown in Fig. 5.37.

Before designing a tiny the tin tensor, there were several requirements imposed:

1. long-term autonomous work,
2. reliability of data sending and receiving,
3. a large communication range,
4. small size and weight,
5. resistance to climate influence.

During preparing tiny biosensors to serial production, all the requirements were complied.

In most cases, the efficiency of data acquisition system work is defined by level of used technologies of data acquisition and data transmission medium. With increasing system scale, the contribution of these components in total efficiency of data acquisition system increases too. So, first of all, it is important to choose a technology of data transmission. Analysis of data transmission technologies showed that it is the most optimal to create such a data acquisition system on the basis of a wireless network with mobile terminals, which use existing mobile communication systems. In most cases it is unnecessary to issue the license on radiofrequency channel using and buy expensive transmission and receiving equipments.

In developed data acquisition system smart portable devices are used as mobile tools for data acquisition in any country region. Measured data from portable devices are transmitted to central control station, where these data are processed, analyzed and generalized. Collected results are offered in graphic, tabular or another form and then used for building generalized map of some region state. Map of state defines, for example, plant cover state of agricultural lands, ecological state, existence of plant diseases etc. on some territory or the whole country.

In our opinion, the main disadvantage of such a system is the following: if you want to change or implement new applied calculation methods, you need to reprogram all the mobile portable devices "Floratest".

Example of the structure organization of the developed data acquisition system is shown in Fig. 5.38.

To remove this disadvantage, we are developing new data acquisition system, which has significant differences from the previous one. Principle of such a system operation is as following. The portable devices "Floratest" measure plant state without any data processing. Then user defines the examined medium (e.g. plant sort) and "raw data" are transmitted to

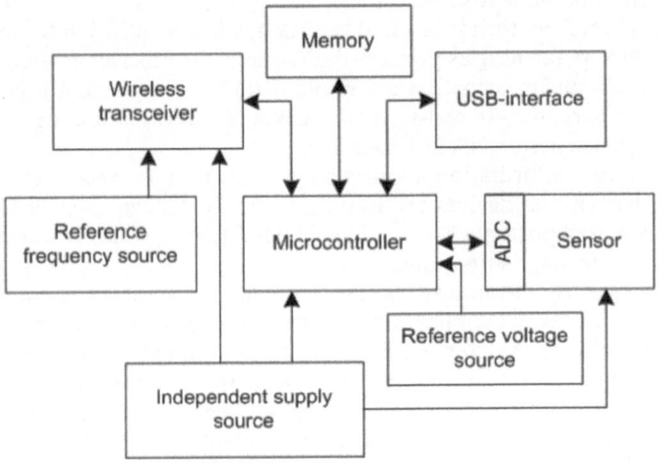

Figure 5.37. Functional scheme of the tiny biosensor of the "Floratest" family.

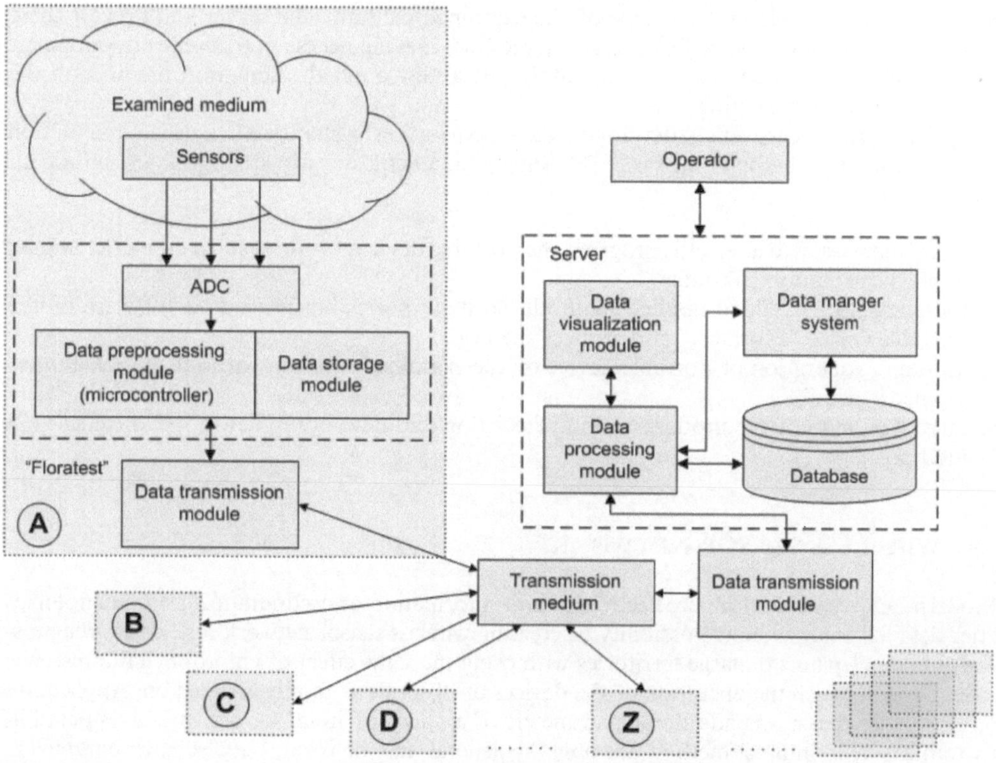

Figure 5.38. Data acquisition system on the basis of the device "Floratest" family.

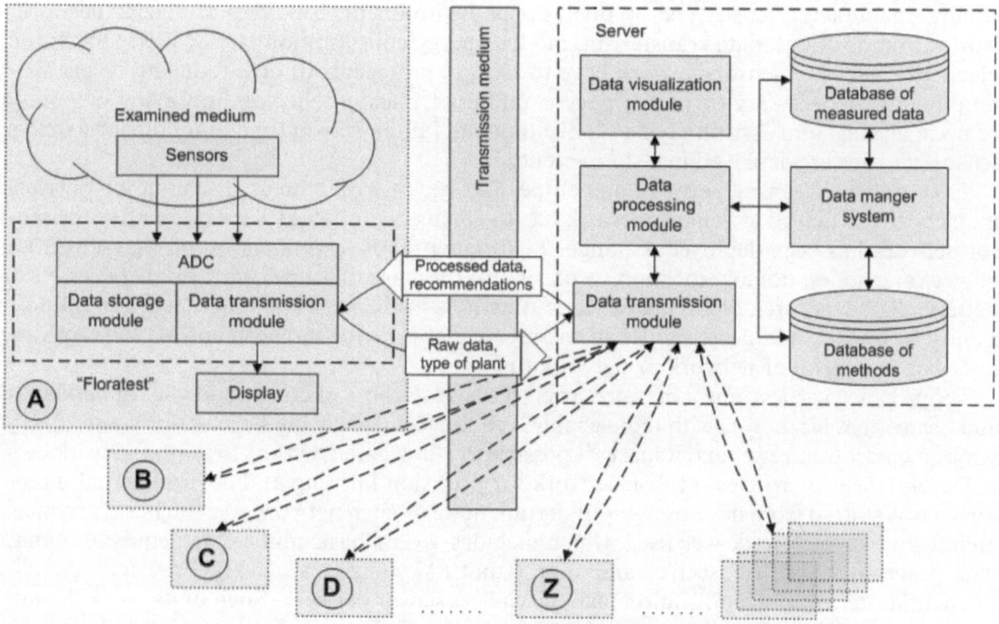

Figure 5.39. Structure of the data acquisition system with applied calculation methods on server.

server. Taking into account the type of the examined medium, the server will process these "raw data" by means of specific calculation method. Having necessity to alter or develop new applied calculation methods, we will simply put a new applied calculation method on the server or replace the existing one.

Typical structure of such developed data acquisition system with applied calculation methods on server is shown in Fig. 5.39. Such a data acquisition system gives us significant advantages:

1. it is unnecessary to alter or reprogram the portable devices "Floratest" in case of changing applied calculation methods;
2. there are no embedded applied methods, so it simplifies hardware and software of the portable device "Floratest" and reduces the price;
3. automatic operation of portable device on the examined medium without specialist interference is possible;
4. large flexibility of the modernization of existing and developing new applied calculation methods.

5.9 WIRELESS SENSOR NETWORK

Modern achievements in microelectronics and information and communication technologies after detailed analysis show possibility of creating wireless sensor network for express-diagnostics of state of plants on large territories with using the same effect of chlorophyll fluorescence induction, on which the operation of the devices of "Floratest" family is based on. Application of sensors for measuring additional parameters of air and soil in sensor network gives possibility to measure in indirect method not only the general state of plant, but also such parameters as pesticides content in soil, water and plants, level of soil corrosion, level of pollution in air etc.

Wireless Sensor Network (WSN) applied for our problem has to meet requirements of typical sensor network for application in agriculture and environmental protection. Development of wireless sensor network depends on many factors, including failure-resistance, scaling, manufacture cost, type of operational environment, topology of sensor network, hardware restrictions, data transferring model, energy consumption etc. Schemes for initial placing of nodes of sensor network have to meet requirements of cost reducing of installation, removing necessary of preliminary planning of placing schemes, improving flexibility of node placing, and assisting self-organization and failure-resistance. Nodes of our wireless sensor network are deployed in a static mode.

Since nodes of sensor network sometimes fail because of battery discharge or physical impacts, the structure of sensor network has to change its topology very often after the sensor network has been deployed. Changes of topology cause modification of characteristics of network nodes, notably location, accessibility (because of noise, moving obstacles etc.), battery charge, failures, changing of node functions in the network. There is specified possibility to include additional nodes in the sensor network for replacement of non-working nodes or expansion of network or network functions.

The proposed wireless sensor network covers the large territory of crop, woodlands or parklands and microelectronic sensors with radio channel are situated directly on the leaves of plants. Every sensor includes transceiver and it makes it possible to join separate sensors to sensor network.

Development of wireless sensor network for precision farming and environmental monitoring was started from development of certain node of such network. As the basis, a typical architecture of such node was used, which includes several basic units: data acquisition unit, data processing unit, transceiver and supply unit (Akyildiz et al. 2002). Presence of additional units depends on application task of wireless sensor network. Such units, as GPS unit, power generator and mobilizer can be considered as additional elements. Data acquisition unit, as a rule, includes sensor and ADC.

The functional scheme of network node is shown in Fig. 5.40 and node itself is shown in Fig. 5.41.

According to the functional diagram of the wireless node, the sensor includes proper optical sensor to convert induction of chlorophyll fluorescence into an electrical signal, microelectronic circuit of microcontroller JN5168, keeping ADC, proper microcontroller and transceiver, and power supply. In this microcontroller electrical signal is converted into the digital code. The code is formatted to passing it to remote mobile computer.

The data processing from every node is executed with network coordinator. To perform additional calculation, the network coordinator includes separate microcontroller. According to analysis of performances of microcontrollers we selected 32-bit floating-point microcontroller Cortex-M4 LPC4357 of Cortex-M family, designed by ARM Company, and manufactured by NXP Company. In addition to this, microcontroller includes analog-to-digital and digital-to-analog converters for input/output analog signals.

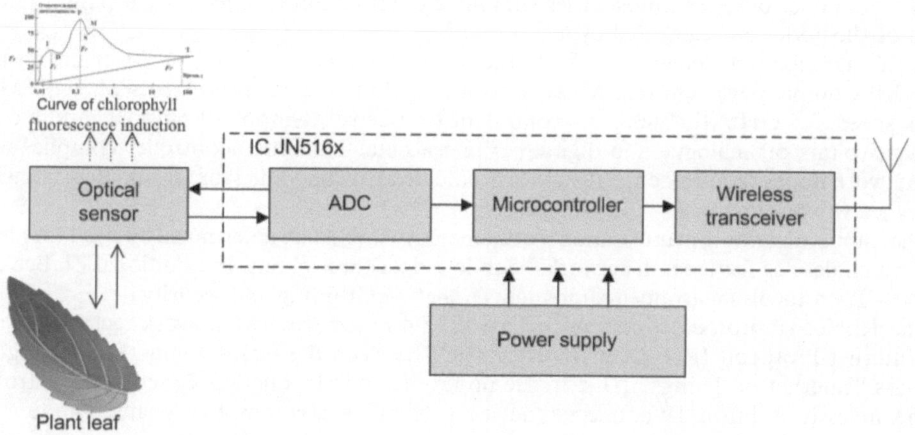

Figure 5.40. Functional scheme of wireless node.

Figure 5.41. Node of wireless sensor network.

Designing WSN engineers have to pay attention to important aspects: in-network data handling and organization of communication between network nodes. Data handling includes development of sensor interfaces, data acquisition, data compression, and data aggregation, data processing, decision-making and report forming. Organization of communication between network nodes includes adjustment of wireless modules, network synchronization, planning MAC-protocols, topology control, strategy of network relations, and common organization of data dissemination.

On a physical layer according to the IEEE 802.15.4 protocol, there are three frequency bands for WSN operation: channel in 868 MHz frequency band, ten channels in 915 MHz frequency band and sixteen channels in 2,4 GHz frequency band for industrial and medical purposes. As a rule, the minimum required power for transmitting a signal over a distance d, is proportional to dn. The variable n is between 2 and 4 and depends on the transmission environment. In most cases, the antenna in WSNs is placed near ground and thus the variable *n* normally equals 4 (Pottie and Kaiser 2000). For bands with higher frequencies, this variable n mainly depends on possibility of line-of-sight path. Usage of signal multi-hop relay from node to node in most cases simplifies the problem of line-of-sight path, because most of the nodes are separated by a short path.

For designing pilot elements of WSN, the wireless microcontroller JN5168 manufactured by NXP Company was chosen. Microcontroller includes 32-bit processor with 1–32 MHz clock speed, 2.4 GHz IEEE802.15.4 compliant transceiver, 4-input 10-bit ADC and a comprehensive mix of analogue and digital peripherals etc. This microcontroller complies with the network nodes requirements. It supports data acquisition, data storage and data transferring via a wireless channel.

The standardization and unification requirements for WSN are determined by 802.15.4 standard, which defines features of networks with low data throughput. In additional ZigBee and JenNet-IP, protocols determine requirements to network routing and security.

The JenNet-IP protocol combines IEEE802.15.4-based wireless network technology and the Internet Protocol (IP) to achieve integration between the two domains, supporting the wireless "Internet of Things". Due to the nodes of a wireless network are to be controlled both wirelessly and from IP-connected device, such as a smart remotely located phone.

Topology control layer supports the determination of data transfer energy consumption for certain units and type of communication of such units. As a rule, the minimum transmission delay in the network has to be reached for minimum energy consumption. Optimal selection of in-network data transfer protocol lets to get optimal ratio of above mentioned parameters.

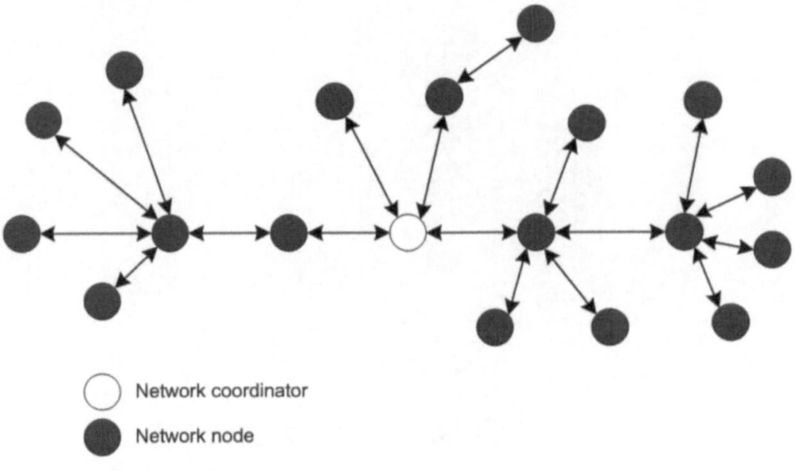

Figure 5.42. Typical architecture of the sensor network.

On the initial development stages for proposed tasks, using of a simple tree-like topology of WSN was proposed. The part of WSN topology, which was created as a base fragment (cluster) of the network, is shown in Fig. 5.42.

5.10 MULTILEVEL SENSOR NETWORK

It is important to note that proposed WSN is planned to be a part of a multilevel sensor network, which operates with usage of "cloud" technologies. The multilevel sensor network is intended for a prompt diagnose a of plant state, knowledge base filling and managerial decision making. The network includes following elements: wireless sensor network as primary data acquisition system, knowledge bases of current and normal (reference) states of plants, processor for managerial decision making and user interface.

The functional diagram of the proposed multilevel sensor network is shown in Fig. 5.43. Knowledge base contains information on state of different plants in normal climatic conditions, acceptable deviations of these conditions, and means for compensation of influence of external stress factors and etc. Knowledge base is filled in a real-time mode with information of current changes of state of plants or trees that were caused by stress factors of natural or anthropogenic origin. Information of a plant state is formed by means of the wireless sensor network. Nodes of this wireless sensor network are placed directly on the plant or tree leaves.

Processor of multilevel network works in "cloud" environment, which improves the results of managerial decision making during fast changes of knowledge of current state of plant under climatic changes and anthropogenic impacts. The processor operation provides handling of different agrotechnical situations and actions for decision making (Kryvonos 2009). User interface is developed as web-interface, what ensures the interaction with different types of data and knowledge and gives the possibility to adopt such an interface with the user requirements.

So plant state data are passed from wireless sensor network to the next level of the multilevel network. The managerial decision "to water or not to water", or "to fertilize or not to fertilize", "to add herbicide or not" and so on is determined with data processing. It is expected that database is permanently replenished with new data of stress factor influence on the plant state, the processors of cloud system generates a managerial decision in real time and user of Web-interface supports the filling of database from different information providers.

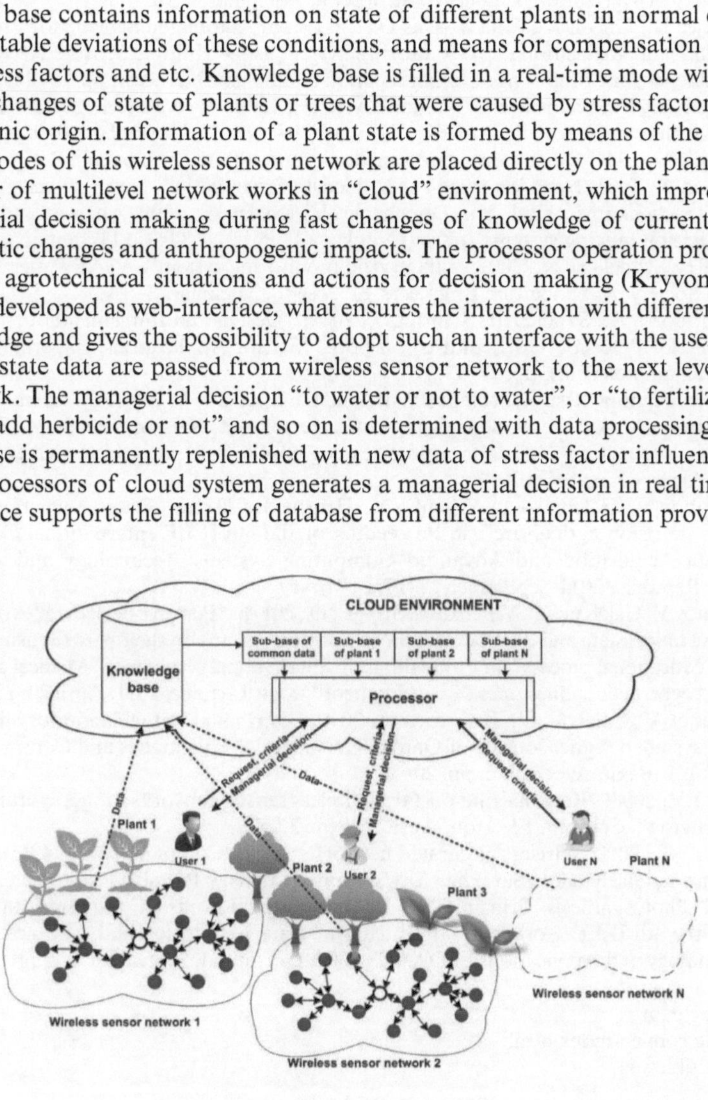

Figure 5.43. Functional diagram of the multilevel sensor network for precision agriculture and environmental monitoring.

REFERENCES

Akyildiz I.F., Su W., Sankarasubramaniam Y., Cayirci E., 2002: "Wireless sensor networks: A survey", in "Computer Networks", No 38, pp. 393–422.

Bulatov M.I., Kalinkin I.P., 1972: Practical manual for photocolorimetric and spectralphotocolorimetric methods of analysis. Leningrad: Himiya.

Camilli A., Cugnasca C.E., Saraiva A.M., Hirakawa A.R., Corrêa L.P., 2007: "From wireless sensor to field mapping: Anatomy of an application for precision agriculture", Comput. Electron. Agric. 58, pp. 25–36.

Evans R., Bergman J., 2007: "Relationships Between Cropping Sequences and Irrigation Frequency under Self-Propelled Irrigation Systems in the Northern Great Plains (Ngp)", USDA Annual Report. Project Number: 5436-13210-003-02. June 11, 2003–December 31.

Gomide R.L., Inamasu R.Y., Queiroz D.M., Mantovani E.C., Santos W.F., 2001: "An automatic data acquisition and control mobile laboratory network for crop production systems data management and spatial variability studies in the Brazilian center-west region", ASAE Paper No.: 01–1046. The American Society of Agriculture Engineers, St. Joseph, Michigan, USA.

Haldar N., Banerjee D., Ghosh K., Jana S., Das D., 2011: "An Automated Scheme for Precision Agriculture through Data Acquisition and Monitoring System using Multiple Sensors Network", in 2nd National Conference on Computing, Communication and Sensor Network, pp. 19–24.

Jeong Gil Ko, Chenyang Lu, Srivastava M.B., Stankovic J.A., Terzis A., Welsh M., 2010: Wireless Sensor Networks for Healthcare/Proceedings of the IEEE. Vol. 98, No. 11, November 2010. pp. 1947–1960.

Kautsky H., Hirsch A., 1934: Das Fluoreszenzverhalten gruner Pflanzen. Biochem Z, pp. 422–434.

Kautsky H., Hirsch A., 1931: Neue Versuche zur Kohlenstoffassimilation. Natur wissenenschaften.

Kryvonos Yu., Kyrychenko M., Krak Yu., Donchenko V., Kulyas A., 2009: Analysis and synthesis of situations in decision making systems. – Kiev: Naukova Dumka, p. 334 (in Ukrainian).

Lee W.S., Burks T.F., Schueller J.K., 2002: "Silage yield monitoring system", ASAE Paper No.:02–1165. The American Society of Agriculture Engineers, St. Joseph, Michigan, USA.

López J., Soto F., Iborra A., Sánchez P., Suardíaz J., 2009: "Design and Implementation of a Wireless Sensor Network for Precision Horticulture", in book "Sensor Applications, Experimentation, and Logistics", N. Komninos (Ed.), pp. 27–42.

Measurement of leaf/canopy level Reflectance With The UniSpec-SC. WWW/PPSYstems.cjm.

Morais R., Fernandes M.A., Matos S.G., Serodio C., Ferreira P., Reis M., 2008: "A ZigBee multi-powered wireless acquisition device for remote sensing applications in precision viticulture", Comput. Electron. Agric. 62, pp. 94–106.

Palagin O., Romanov V., Galelyuka I., Velichko V., Hrusha V., 2011a: "Data acquisition systems of plants' state in precision agriculture", in Proceeding of the 6th IEEE International Conference on Intelligent Data Acquisition and Advanced Computing Systems: Technology and Applications, Prague, Czech Republic, 2011 September 15–17, pp. 16–19.

Palagin O., Romanov V., Galelyuka I., Velichko V., Strygak O., 2011b: "Portable devices for express-diagnostics of biological object state and distributed data acquisition systems on their base for using in scientific researches and educational process", in Proceedings of International conference "Medical and biological informatics and cybernetics: milestones of development", Kyiv, Ukraine, 2011, April 20–23, p. 28.

Palagin O., Romanov V., Galelyuka I., Kovyrova O., 2010: Distributed virtual laboratory for smart sensor device and system design // Journal of Qafqaz University "Mathematics and Computer Science". Number 29, Vol. 1, Baku, Azerbaijan, pp. 20–36.

Pierce F.J., Elliot T.V., 2008: "Regional and on-farm wireless sensor networks for agricultural systems in Eastern Washington", Comput. Electron. Agric. 61, pp. 32–43.

Pottie G.J., Kaiser W.J., 2000: Wireless integrated network sensors. Commun. ACM, 43(5): pp. 51–58.

Romanov V., Fedak V., Galelyuka I., Sarakhan Ye., Skrypnyk O., 2007: Portable Fluorometer for Express-Diagnostics of Photosynthesis: Principles of Operation and Results of Experimental Researches. Proceeding of the 4th IEEE Workshop on "Intelligent Data Acquisition and Advanced Computing Systems: Technology and Applications", IDAACS'2007, Dortmund, September 6–8, pp. 570–573.

http://www.foryard.com.

http://www.taosinc.com.

http://www.analog.com/en/index.html.

http://www.vdmais.kiev.ua.

CHAPTER 6

The mathematical problems of complex systems investigation under uncertainties

K. Atoyev, P. Knopov & V. Pepeliaev
V.M. Glushkov Institute of Cybernetic of NAS of Ukraine, Kyiv, Ukraine

P. Kisała
Lublin University of Technology, Lublin, Poland

R. Romaniuk
Warsaw University of Technology, Warsaw, Poland

M. Kalimoldayev
Institute of Information and Computational Technologies, Almaty, Kazakhstan

INTRODUCTION

The threats that will be most challenging in the future are connected with synergistic interactions between risks of various origins. There are constellations of interconnected risks within a community described as a "risk envelope" (Wilkinson et al. 2003). That is why our regulatory focus moves from managing point sources and risk in isolation to one of managing and trading risks from multiples sources and interdependencies of ecological, economic, social and technogenic risks.

The cumulative effects of various stressors emanating from multiple sources become important dangerous generators of uncertainty and instability in modern society. Indeed, interdependencies of various risks increase the difficulty of identifying the principal cause of injurious effects and make system already weakened by stress more susceptible to the effects of additional stressors.

One of the most significant indicators of system's security is adaptiveness to various impacts, the ability of a system to change their internal structure in convenience with environmental changes due to mutation or activation of internal systemic transformations. The main forces driving these transformations are dependent on the energy support.

The stress response is also associated with the mobilization of energy to enable the system to deal with threat or danger. However, energy cannot be diverted in all directions simultaneously, so in the case of cumulative stressors takes place the nonlinear increasing of system's vulnerability and decreasing of its adaptive possibilities. These nonlinearities lead to arising of areas of system parameters values that are often characterized by unpredictable behaviors, when small impacts cause essential transformation of system's security.

The rise of uncertainty and unpredictability is often connected with regimes when system shows sudden, discontinuous changes or phase transitions as a result of small, continuous changes in variables that influence the system.

Another reason of the essential increasing of rare events is caused by novel form of global challenges, resulting from wide range of asymmetric threats, high susceptibility to initial conditions, strengthening destroy power of rare events, increased possibilities that some minor disturbances can cascade into a disaster with unpredictable ecological, economic and political consequences.

The deformation of security's area causes the crisis of control systems and decreases the horizon of forecasting for extreme events. The modern society can be considered as an open, non-Hamiltonian, non-linear, dissipative, potential, self-reproducing, self-organizing system with networking elements, near an unstable steady state far from equilibrium, in bifurcation zone where any fluctuation leads to an unpredictable future.

Traditional methods of risk analysis often are not enough efficient for such events when system is very susceptible to initial conditions and its behavior shows sudden, discontinuous changes or phase transitions as a result of small, continuous changes in variables that influence the system. Indeed, as pointed in (Wilkinson et al. 2003, p. 388), "risk management tools designed on the assumption that historical variability and existing trends are relevant to future context are likely to be increasingly challenged by growing interdependencies of environmental, technological, social and economic hazards". The unique and single character of rare events makes it difficult to use correctly the theory of probability for proper risk assessment.

Moreover, it should be noted that traditional model-based assessments have unavoidable uncertainties linked to underlying model assumptions and available input data. Furthermore, globalization leads to more complex and interconnected network of modernized society, including its ecological component. The behavior of such complex system sometimes becomes unstable and shows sudden, discontinuous changes or phase transitions as a result of small, continuous changes in variables that influence the system. Small causes can lead to large consequences and generate chaotic oscillations of vulnerability, since the amplifications of global change play important role in appearance of the supported chaotic regimes of environmental variables.

How can we decrease the degree of uncertainty in forecasting of environmental dynamics? The "integrated" global models for sustainable management of the environment are not efficient for systems with high susceptibility to initial conditions, for which some minor disturbances can cascade into a disaster with unpredictable ecological consequences. The new risks demand new methods for their assessment.

In (Atoyev 2015, Atoyev 2007, Atoyev 2001, Atoyev, Tomin and Aksionova 2007) a decision support for investigation of interconnected risks with the help of catastrophes theory, theory of chaos and bifurcation is proposed. It increases our possibilities in the field of systemic risk analysis and management. The main advantage of this approach is a determination of risk as a function of dynamic variables of the investigated system.

The developed approach may be applied to a wide range of different complex systems from following disciplines:

1. ecology (assessment of environmental vulnerability, global change impact, reducing water stress and optimal water resource management);
2. epidemiology (estimation of HIV/AIDS sharp escalation for management of epidemics and risk ranking);
3. biology (healthcare, cellular energetic trigger, metabolic and hormonal regulation);
4. psychology (formation of attitude, mutual penetration of west and orient worldview matrices and terrorism);
5. sociology (deformation of social risk's fields).

6.1 GLOBAL CHANGE AS MAIN GENERATORS OF NEW RISKS

6.1.1 *The main reasons of reducing efficiency of forecasting under the global changes*

The disaster on the Japanese nuclear power plant Fukushima has focused attention on the crisis of modern control systems. This crisis is manifested in reduction of forecasting horizon and impossibility of rare events prediction, such as the Chernobyl disaster, the hurricane Katrina, the Deepwater Horizon disaster in the Gulf of Mexico in 2010, the synergy effect of earthquake and hurricane in Japan in 2011, the world financial crisis, and the geopolitical destabilization in some Muslim countries in 2012.

The frequency of emergencies and the scale of corresponding losses appeared essentially larger than it was predicted by forecasting methods based on the traditional approaches of risk analysis. Reducing efficiency of modern forecasting and risk control systems is explained, first of all, by the global changes causing the general instability of contemporary World. The unique, accidental character of turbo-events passes away while non-homogeneity and non-uniformity become an ordinary attribute of information society. The turbulent character of life causes new risks. That is why our times were named as "epoch of bifurcation time" (Laszlo 1993).

Creating a complex network of modern society strengthens the fields of technogenic, economic, and environmental risks. It is explained by such features of modern society as the increased density of links among network nodes and the increased speed of circulations of tangible and intangible assets throughout the network. The network nodes can play the role of special acupunctural points of society vulnerability, in which small disturbances may lead to essential divergences of system development trajectories, unpredictable in their scale consequences, increases of future uncertainty. The functioning of such complex systems is becoming unstable and characterizing by discontinuous changes in their behavior. The blackouts of power systems in the USA and Western Europe (2003–2011) give examples of that instability of technogenic systems, whereas the financial crisis and political crisis in the Arab World in 2012 provide examples of the instability of social-economic systems. Therefore, a network society can be viewed as a system around the fragile steady state in bifurcation zone under increasing probability of rare and unpredictable events. Very small fluctuations may lead to sudden discontinuous changes of social-economic and environmental-technogenic characteristics of society.

The common feature of such systems is a strong dependence on initial conditions. The increasing role of small impacts extends the action areas of traditional risks and generates the principally new risks. These risks are related with the events of tremendous destructive power, which are characterized by their large duration and transboundary impact when the consequences of technogenic and natural disasters are visible in the territories which are located far away from the emergency source. Research of system behavior around the fragile steady state in bifurcation zone is facing the principal difficulties reducing the efficiency of forecasting.

The first reason of those difficulties is due to a number of new risks. Tools of probability theory are efficient if representative samples are available. The higher degree of risk novelty, the lower quality of decision making as a unique character of event presumes absence or incompleteness of statistical data. New methods for risk estimation should be developed in order to take into account the new risk nature under the global changes.

The second reason is due to power law of probability density function for losses and causalities under natural disasters during global changes. Large deviations (from the expected value for losses and causalities) can be neglected if probability density function is Gaussian (normal). But many disasters generate statistics described by power law probability distribution where large deviations cannot be neglected. The power law distribution has the following feature: the losses at only event may exceed the sum of losses at the rest of events within the sample.

During the past decade, we have witnessed novel form of global challenges, resulting from wide range of asymmetric threats, strengthening destroy power of a renewed focus on events, increased possibilities that some minor disturbances can cascade into a disaster with unpredictable ecological, economic and political consequences. Short list of different generators of new risks in web society includes following items:

1. increasing role of social amplification of threats;
2. existence of longer-term links and a great extent of causal chains;
3. increasing role of systemic risks;
4. increasing geographic concentration of wealth, human capital, knowledge and communication links;
5. progress in communications and information processing.

Uncertainty and principle unpredictability has become the keywords characterizing various aspects of global change. The deformation of global security causes the crisis of control systems, leads to sharp increasing of environmental and man-made disasters, financial and economic catastrophes and decreases the horizon of forecasting for extreme events, especially when probability density function of destroy power is described by power low. Its curve decreases more slowly than Gaussian probability low (Fig. 6.1).

The GC (as pointed by F. Capra (2002)), can be considered as a giant web, which consists of fractal networks where any network contains another one. The quickly changing world with its instability and jumps to new forms of the organizations is a good example of Prigogine's theory of dissipative structures (Prigogine and Stengers 1984). These instabilities appear to be a result of fluctuations amplified by positive long-term feedback loops. The amplifications of GC play important role in appearance of the supported chaotic environmental and social regimes of global flows.

GC plays an important role in arising of new risk. One of the typology of new risks has been proposed by O. Renn (2003). These risks fall into five main categories: 1) psychosomatic risks, 2) system breakdown risks, 3) systemic risks, 4) knowledge management risks, 5) global change risks. The others new taxonomies of risks also include risks resulting from: 1) homogenization of infectious diseases; 2) occurrence of a new pathogens (AIDS, Severe Acute Respiratory Syndrome (SARS), Bovine Spongiform Encephalopathy (BSE) or mad cow disease); 3) over–prescription of antibiotics; 4) advances of functional genomics, DNA technologies, genetic and protein engineering and genetically modified food and insidious risks, which emerge slowly but surely over time (Wilkinson et al. 2003, p. 386); 5) terrorist threat; 6) mental inadaptation of personality connected with atomization of society and increased social volatility. The hacker-type risks, and transboundary risks also are included in new taxonomies of risks.

Climate change is another enormous global threat. Various scenarios of Intergovernmental Panel of Climate Changes indicate some possibility of irreversible catastrophic effects that could wipe a great part of the world population due to global warming. Moreover, the climate changes cause the disturbances of traditional biogeocoenosis that change the routes of transmigration of birds of passage. One of possible consequences of these processes is expansion of the avian influenza ("bird flu").

The next group of new risk is subjective risk associated with prospect theory. One very important result of Kahneman and Tversky work is demonstrating that people's attitudes toward risks concerning gains may be quite different from their attitudes toward risks concerning losses (Kahnemann and Tversky 1972). A. Boholm has proposed to separate this phenomenon as the subjective component of risk. "Subjective risk acknowledges that people's beliefs and opinions tend to deviate from such scientific assessments. People understand and judge risks in terms of locally defined values and concerns" (Wilkinson et al. 2003, p. 371).

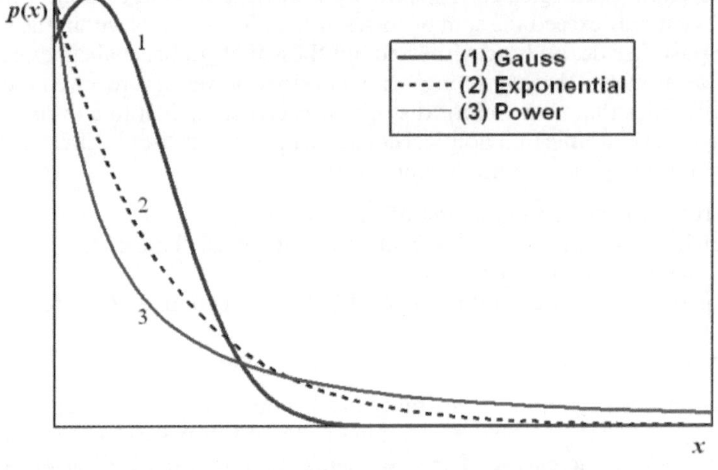

Figure 6.1. The typical view of Gauss (1), exponential (2) and power (3) lows.

Subjective risk can be examined as an additional powerful source of sharp escalation of man-made mistakes and unpredictability of human behavior which decrease security.

So, the transformation of security is carried out in space with principle new features caused by globalization and increasing uncertainty. The changing of fundamental characteristics of time-space continuum generates the fields of new risks, leads to more unpredictable world, increases the degree of global mega threats (atomic, genetic, chemical, mental, environmental) and their damage. Moreover, the revolution in quality of communication links, necessity of fast solutions and increasing role of psychological factors for decision making under the face of uncertainty also transform area of security.

New risks demand new methods for their assessment. That is why the theories of defense planning, psychological and synergetic economics, theory of catastrophes took one of the central places for risk analysis in era of uncertainty (Klinke and Renn 2002). Adaptiveness describes the ability of a system to transform their internal structure in response on environmental changes. It has become one of the most significant characteristics of sustainable development of system.

The main objective of current investigations is to strengthen the capacity in such fields as impact of uncertainty caused by Global Change (GC) on transformation of security and sustainability of complex systems for which standard modeling methods and tools are inadequate. The work includes following tasks: 1) to develop the advanced modeling method to investigate regimes with sudden, discontinuous changes or phase transitions from one level of security to another; 2) to determine the complex system's vulnerability to small fluctuations; 3) to calculate the optimal controls, minimizing the decreasing of security; 4) to investigate the possible mechanisms of chaotic behavior arising that cause the instability of complex systems.

6.1.2 *Development of the advanced modeling method to investigate regimes with sudden, discontinuous changes or phase transitions of complex system's security*

In (Atoyev 2007, Atoyev, Tomin and Aksionova 2007) it was developed the modeling method for risk assessment on the basis of mathematical formalism for modeling nonlinear systems whose discontinuous behavior is determined by smooth changes in a few driving parameters—catastrophe theory (CT) (Thom 1975). CT has been applied to a wide range of different systems from physics, engineering, biology, psychology, and sociology. A small subset of specific phenomena that were analyzed and modeled using CT includes quantum morphogenesis, the formation of caustics in ray optics, the stability of ships, the stability of black holes, the surface changes in interatomic potential, Euler struts, the size of bee societies, morphogenesis or cell differentiation in embryology, bistability of perception, binocular vision, motor learning, sudden transitions in attitudes, and the cognitive development of children (Poston and Stewart 1978). In all these applications, the behavior of the system under study shows sudden, discontinuous changes or phase transitions as a result of small, continuous changes in variables that influence the system (cf. the freezing of water when temperature is gradually decreased or the collapse of a bridge under slowly mounting pressure). In general, CT applies to systems that may respond to continuous changes in control variables by a discontinuous change from one equilibrium state to another.

Let us propose that complex system satisfies all the requirements of potential system and can be described by potential function $U(x, A_1, A_2, \ldots, A_N)$ of the behavioral variable x and control variables A_i. The dynamics of deterministic gradient system are described by a differential equation of the form:

$$dx / dt = -\partial U / \partial x. \tag{6.1}$$

Equation (6.1) implies that variable x changes in the direction of decreasing potential at a rate proportional to the slope or gradient of the potential field. The equilibrium manifold of this system is the set of values x such that $dx / dt = -\partial U / \partial x$. For example, if:

$$U(x, A_1, A_2, A_3, A_4, A_5) = 1/6x^6 + 1/4A_1x^4 + 1/3A_2x^3 + 1/2A_3x^2 + A_4x + A_5, \qquad (6.2)$$

then the system is in equilibrium, when $x^5 + A_1x^3 + A_2x^2 + A_3x + A_4 = 0$.

Function *U* has 5 steady states; 3 of them are stable, 2 states are unstable. The transformations of system from one steady state to another, or transformation of a character of a steady state (for instance, from stable to unstable) are functions of control parameters *A*. These parameters control both a movement of system's current position upon surface *U*, and transformations of the surface itself.

The probability of sudden and unexpected transformation from one steady state to another may be determined using the theory of catastrophes allowing the calculation of bifurcation values, curves and surfaces of control parameters. The probability of transformation is estimated as an extent of a control parameter approaching these bifurcation values characterizing system's transition from one steady state (norm) to another (catastrophe).

It is possible to determine the quantity of steady states of system on the basis of initial data set analysis. Let us propose that this analysis shows that investigated complex system has three stable steady states. To make it simple, it is possible to put forward following suppositions. The first stable steady state characterizes normal condition. The level of x in this state is minimal. The second one characterizes state with a medium level of x. The third one characterizes crisis with high level of x. Under this supposition, the potential function U is described by equation (6.2).

There are many ways in which society is changing. Four key forces that are driving GC were identified in (Wilkinson et al. 2003). They are increasing global connectivity, the atomization of society, rapidly advancing technology and growing ability to measure and monitor. To examine a GC impact problem, we also have chosen a 4-dimensional typology. The transformations of security are defined by one of the universal deformations of catastrophes' theory—a "butterfly":

$$-\partial U(X, A) / \partial X = X^5 + A_1X^3 + A_2X^2 + A_3X + A_4, \qquad (6.3)$$

where A_i $(i = 1{,}4)$ are control parameters, characterizing various aspects of the GC impacts on economic, social, environmental and technogenic dimensions of sustainable development: $i = 1$ – index of global connectivity (economic index), determined by the powerful forces of economic globalization, market liberalization and economic possibilities to prevent and minimize consequences of disaster; $i = 2$ – index of atomization of society (social index), determined by factors that provoke the rise of violence, mental disadaptation of personality, probability of epidemics and reinforce the notions of individual ability, identity, freedom and responsibility; $i = 3$ – index of ability to measure and monitor (index of natural disasters), determined by ecological and climatic factors that impact on probability of natural disaster and man's growing ability to observe and analyze his environment, $i = 4$ – index of advancing technologies (index of man-made disasters), determined by factors that impact on development of technologies, educing of their lifespan and increasing of their disasters.

The projections of bifurcation surfaces on planes of natural and man-made disasters A_3A_4 are presented in Fig. 6.2. Changes of each variable characterizing parameters A_i are used as control effects. The characterizing point, corresponding to current state of system (circle) had begun moving from small level of natural disaster's index (parameter A_3) under constant levels of parameters A_1 and A_2 (Fig. 6.2a).

The increasing of A_3 decreases distance to bifurcation surface that separates norm and pre-catastrophe (Fig. 6.2b). A small impacts connected with increasing of man-made disaster's index (parameter A_4) when system parameters balance at the boundaries of their bifurcation values lead to transformation the system state from norm to pre-catastrophe (Fig. 6.3c). Further increasing of stressors leads to new transformation from pre-catastrophe to catastrophe (Fig. 6.2d).

The risk estimation of sudden and unexpected transformation from one steady state to another may be performed using the theory of catastrophes allowing the calculation of

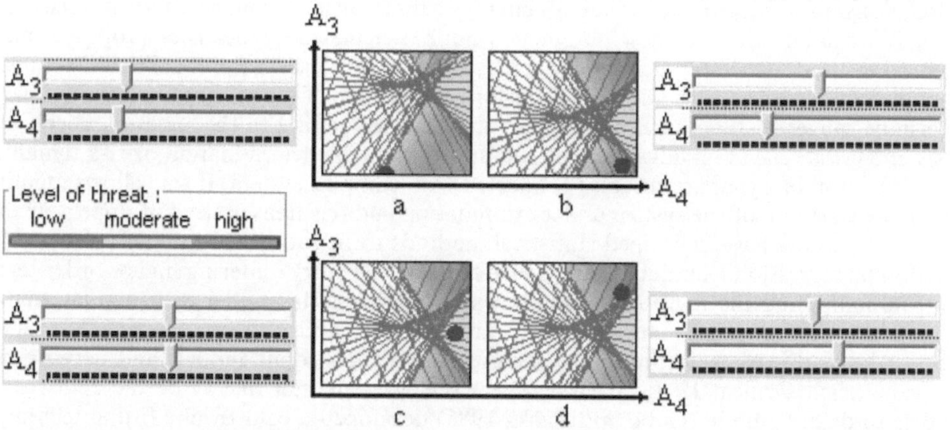

Figure 6.2. Projections of catastrophe on the surfaces A_3A_4.

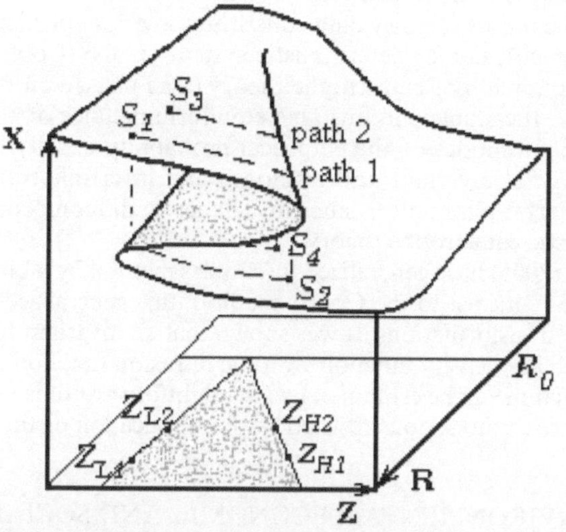

Figure 6.3. Cusp catastrophe model for risk assessment.

bifurcation values, curves and surfaces of control parameters. The risk is estimated as an extent of a control parameter approaching these bifurcation values characterizing system's transition from one steady state (norm) to another (catastrophe). The main advantage of this approach is a determination of risk as a function of dynamic variables of the investigated system. It also allows identifying the weakest link of the system under examination and the areas in need of an improvement.

The following algorithm of risk assessment can be proposed:

1. The collection of information characterizing the above-mentioned indices input from contemporary databases.
2. Determination of indices characterizing appropriate group of parameters calculation by means of developed mathematical models using inputted data.
3. Calculation of the bifurcation parameters values at which the number of system steady states change.
4. Estimation of restoration possibilities of each considered systems by the remoteness of the parameter characterizing appropriate index from its bifurcation value.

Relationship shifts within complex system may cause transfers from norm to pre-crisis or to crisis. Bifurcation values of these parameters can be calculated by means of proposed mathematical method. Reaching such critical values abruptly increases the probability of a transfer from one functional state to another. Thus, for a given state of global system, it is possible to determine ranges of functional parameter values corresponding to the states of norm, pre-crisis and crisis. The main advantage of this approach is the determination of risk dynamics as the function of dynamic variables connected with complex system. It also allows identifying the weakest link of the system under examination and the areas in need of improvement.

Several authors have developed statistical methods to fit catastrophe models. Guastello's polynomial regression technique uses reverse hierarchical entry (entering higher order terms first) in computing the regression equation and compares the model with two alternative linear regression models (Guastello 1998). Lange, Oliva, and McDade (Lange et al. 2000) proposed a multivariate methodology for estimating catastrophe models and developed a program called Gemcat II. It is claimed that the program can fit any of the catastrophe models to data. Cobb in (Cobb and Zacks 1985) developed a catastrophe fitting technique that is based on a stochastic interpretation of catastrophe theory. Taking into account all restrictions of above methods that were analyzed in (van der Maas et al. 2003) we choose the modification of Gemcat II to fit catastrophe model.

Fitting a catastrophe model to noisy data constitutes a serious challenge, however, because CT was formulated specifically for deterministic systems. Cobb (Cobb and Zacks 1985) has developed a stochastic version of catastrophe theory (SCT) based on Ito stochastic differential equations. In SCT, the stable and unstable equilibrium states of the system correspond to the modes and the antimodes of the empirical probability density function, respectively. Unfortunately, SCT is not invariant under smooth and invertible transformations of variables, this was an important limitation, since invariance to diffeomorphic transformations is essential in deterministic catastrophe theory.

Wagenmakers et al. (2005) have generalized the Cobb's method by taking into account the Ito transformation rule. In contrast to the Cobb's method, this generalized SCT is unaffected by smooth and invertible transformations. It was shown that an invariant function is obtained by multiplying the probability density function with the diffusion function of the stochastic process. This invariant function can be estimated by a straightforward time series analysis based on level crossings. So, there are no serious restrictions for application of this theory to noisy data.

6.2 COMPLEX ENVIRONMENTAL, ECONOMIC AND SOCIAL SYSTEMS

6.2.1 *Application of catastrophe theory to wide range of scientific disciplines for investigation of complex systems instability*

Various models are currently available to describe the interrelationship among the parameters of complex ecological—economics systems (Atoyev et al. 2013, Korotayev and Tsirel 2010, Magnitskii and Sidorov 2006, Moiseev et al. 1985, Myasnikov 2011, Puu 2000, Wagenmakers et al. 2005, Zhang 2005). Excessive level of detail in these models is the "price" we pay for trying to adopt model to a specific database relevant to the user. It is necessary to select a small number of generalized parameters that would make it possible to investigate the behavior of a system in a control space of limited number of dimensions. Methods of catastrophe theory have been applied for this purpose in (Atoyev 1999, Sergienko et al. 1997). All the parameters are divided into three groups: recreational parameters X that characterize the renewability of natural resources; load parameters Z that characterize the pollution of the environment; and reserve parameters R that depend on the resources of the ecological—economic system.

The interrelationship between these parameters is described by universal deformation model of catastrophe theory (the cusp model):

$$dx/dt = X^3 + (R_0 - R)X + Z = 0, \tag{6.4}$$

where R_0 – is the model parameter.

The model (6.4) has three stationary states. Two of them are stable, with high $S_1(X_1,R_1,Z_1)$ and low $S_2(X_2,R_1,Z_2)$ levels of recreational and ecological safety (Fig. 6.3).

Assume that initially the system is at point S_1. As the level of pollution Z increases (path 1), the value of X declines, i.e., the recreational value and ecological safety decreases. However, it continues to be fairly high, and only slightly differ from X_1. When the system reaches the lower boundary Z_{L1} of the trigger zone, where three stationary states exist, it moves into a zone of increased risk of ecological disaster. Farther increase of Z brings the system closer to the upper boundary Z_{H1} of the trigger zone. When the system crosses this boundary, the stationary state with a high level of recreational possibilities ceases to exist. The recreational potential the system abruptly falls to the level X_2, which corresponds to an ecological disaster. To initiate recovery, the anthropogenic load must be reduced to the level Z_{L1}.

Let us now consider the case when Z starts increasing from the point $S_3(X_1,R_2,Z_1)$, corresponding to lower system reserves. The boundary of the trigger zone is reached and ecological disaster occurs in the case at a lower level of anthropogenic load on the ecosystem, Z_{H2}. The amplitude of fall of the function X is also smacker (path 2). System recovery becomes possible at a higher level Z_{L2} than in the first case.

According to Shturm's theorem, the cubic polynomial has one or three roots. Their number is depended on discriminant $\Delta = 4R^3 + 27Z^2$. When $\Delta < 0$ there are three real roots, when $\Delta > 0$ there is on real root. $\Delta(a, b) = 0$ is the curve of bifurcation.

So the boundaries of trigger zone are described by following equation:

$$Z_L = -\sqrt{4R^3/27}, Z_H = \sqrt{4R^3/27}. \tag{6.5}$$

The measure of risk of an ecological disaster is provided by the proximity of the system parameters to the boundaries of trigger zone. The risk is defined as:

$$RISK = Z/Z_H. \tag{6.6}$$

The variation over time of the anthropogenic load and ecosystem resources needed for calculated the disaster risk by the proposed approach can be obtained using the existing models of ecological—economic systems (Cellier 2008, Marcuk 1986).

The formalization of interrelations in wide range of complex systems allows identifying the functions $U(X,A)$. The classification of control parameters that are utilized for investigation of discontinuous behavior in ecology, epidemiology, biology, psychology and sociology is presented in Tables 6.1 and 6.2.

Table 6.1. Classification of control parameters for biological, ecological and psychological systems.

	Control parameters	
$\partial U/\partial x$	A_1	A_2
$x^3+A_1x+A_2$	Biological systems: probability of pre- and pathological states (Atoyev 1999, Sergienko et al. 1995)	
Type of	Is determined by internal reserves	Is determined by exposure
catastrophe – "cusp"	Ecological systems: determination of critical levels of pollution and internal reserves of ecological systems (Atoyev 2001, Pepeliaev et al. 2010)	
	Is determined by internal reserves	Is determined by level of pollution
	Psychological systems: attitude formation (Anderson 1985)	
	Is determined by emotional assessment of situation from point of view of its importance	Is determined by rational assessment of situation from point of view of gain and loss

Table 6.2. Classification of control parameters for social systems.

	Control parameters		
$\partial U/\partial x$	A_1	A_2	A_3
$x^4+A_1x^2 +A_2x+A_3$	Social systems: investigation of physical, human and social capitals impacts on social security (Atoyev 1999)		
Type of catastrophe - "swallow tail"	Is determined by index of sustainable development characterizing the impact of endogenous factors (economic, ecological, and social) on changing of social security	Is determined by index of globalization characterizing the impact of exogenous factors (shifts in structure of interrelations in global world) on changing of social security	Is determined by index of social capital characterizing the advantage created by a person's location in a structure of relationships in a frame of radius of trust

The classification of control parameters that are utilized for investigation of different complex systems from ecology, epidemiology and biology, to psychology and sociology with the help of "butterfly" catastrophe is presented in Table 6.3.

6.2.2 *Modelling of uncertainty and risk impact on food and economic fields*

The concept of sustainable development becomes a dominant determining the paths of post-industrial society transformation in the XXI century. The main idea of this concept consists in coordination of the changes in different dimensions (economic, environmental, social) of society functioning. The basic postulate of sustainable development, as confirmed by the UN summit "Rio+20", is creation of conditions for social progress aimed to providing certain quality and safety for people lives, with simultaneous preserving environment, resources, and energy due to adoption of advanced technologies.

The achievement of the best social standards of life; the realization of successive policy for harmonization of basic factors influencing on social development; decrease of social inequality, corruption level, infant death rate; increase of life expectancy, political stability, civil liberties, is becoming a measure of success for society. However, substantial deformation of safety space, related with the global changes, becomes a basic challenge to sustainable development. Growing interdependence of the different dimensions of social vulnerability, wide spectrum of local, regional, transnational threats, expansion of dangerous technologies, deficit of water, food and energy resources, constantly growing load on the economy as a result of population aging, contamination of environment create specific synergic effects and bring additional systemic risks.

Sustainable development of society assumes a reasonable balance between the activities, directed to an acceleration of economic growth, and efforts, directed to mitigate a negative impact of such activities on environment. This problem is of great importance in the sphere of agro-industrial production that is the most sensible pin zone in the system of reciprocal influence of economic and ecological interests of humanity. The problem of food security is getting extra importance due to the global climatic changes which substantially affect the agricultural productivity. The adaptation of agriculture to new climatic terms becomes the priority task.

Efficiency of solving the tasks mentioned above depends on the technological base to be implemented for problem solution. In addition to the changes, usually called global challenges, the transformation of technological modes (the transition from information to post-information society) is beginning. Bio- and nano-technologies, robotics, new environmental management, new high humanitarian technologies, related with human endowment and information load minimization for a decision maker, become the top directions of successful development.

Future designing, scientific monitoring, risk management will occupy the leading technological niches. The transformation of technological modes gives a unique chance of shortening

Table 6.3. Classification of control parameters for different complex systems with the help of "butterfly" catastrophe social systems.

Control parameters			
A_1	A_2	A_3	A_4
Global change (Atoyev et al. 2007)			
Is determined by economic possibilities to prevent and minimize consequences of disaster	Is determined by the factors that provoke the rise of violence, mental disadaptation of personality, etc	Is determined by ecological and climatic factors that impact on probability of natural disaster	Is determined by factors that impact on probability of technogenic disaster.
Economic systems (Atoyev 1999, Atoyev 2007)			
Is determined by the factors characterizing the market of labor (index GINI, unemployment, etc)	Is determined by the factors characterizing the market of money (the rate of inflation, level of credits, etc)	Is determined by the factors characterizing the market of valuable papers (Dow Jones index)	Is determined by the factors characterizing the market of goods and services (a level of GNP, wages, etc)
Biological systems (Atoyev 2005)			
Is determined by hormones level, population of lymphocytes, etc	Is determined by the balance between energy synthesis and consumption processes, etc	Is determined by psychical and psychological peculiarities of the organism	Is determined by genetic inclinations of the organism
Epidemiological systems (Atoyev 2002, Lapushenko et al. 2004, Moskalenko et al. 2002)			
Is determined by the factors characterizing the healthcare	Is determined by the factors characterizing the restoration possibility of population	Is determined by the factors characterizing impact of ecological situation	Is determined by the factors characterizing different agents of infectious diseases, etc.
Political systems (Atoyev 2001, Atoyev 2002)			
Is determined by the factors characterizing the nets of ideas	Is determined by the factors characterizing the nets of rules	Is determined by the factors characterizing the nets of actions	Is determined by the factors characterizing the nets of interests

lags behind the leaders to such countries as Ukraine whose indices remain at the levels of pre-information epoch because the leaders are also forced to adopt the transformation. Thus, under decision making on sustainable development, it is necessary to take into account bifurcation character of a current historical moment, corridor of opportunities given by jumping to the post-information technological mode, and catastrophic consequences in the case of inertial scenario, when future development follows old-fashioned technologies.

All this requires elaboration of new methods and mechanisms for efficient prevention of potential risks to national interests and neutralization of available threats. The information systems, allowing solution of the risk management problems and investigation of global change impact on society, must be a core of such mechanisms. Solving this problem is not feasible without development of efficient mathematical tools of risk estimation for agro-industrial production and balanced use of food, water and energy resources under increased uncertainties caused by the global climatic changes.

The efficiency of modelling largely depends on finding a balance between model particularity and simplicity. The more detailed model, the larger its scale and difficulty of input data collection, parameters identification, the more error of calculation, the less horizon of prediction. Moreover, a complex system is not simply a sum of its subsystems. However, without models detailed enough, it is hard to understand how the disturbances are spreading throughout the subsystems, how adaptations of some subsystems are influencing on activities of other subsystems, how self-organization mechanisms are forming complex social-economic system as a whole. Without such models, finding the actual scenarios of development and evaluating the efficiency of particular management decisions are not feasible.

The large-scale problems often may be solved due to non-homogeneity of the model phase space. For some parts of the phase space, the state of system can be characterised satisfactory enough by a small group of variables comprising a small-scale projection. Other variables can be subordinated to the variables of projection (order parameters). Those other variables are not important from the viewpoint of system description within the given area of parameters. In general, the small-scale projections can be used in the bounded areas of phase space, while those projections are different for different areas.

The system behavior is unpredictable, complicated and sophisticated in the areas of phase space, where such small-scale projections and the corresponding methods of nonlinear dynamics are not plausible; then the methods of probability theory are used (Zulpukarov et al. 2006). Therefore, the combination of dynamic (Atoyev et al. 2014, Sergienko et al. 1997) and stochastic (Atoyev et al. 2014, Ermoliev and von Vinterfeldt 2012, Knopov and Derieva 2013, Knopov and Korkhin 2012, Knopov and Pardalos 2009) models is applied to research of complex systems. The system behavior is investigated with support of the methods smooth and multi-valued mappings (Atoyev et al. 2007, Sergienko et al. 1997) in the cases when the probability methods become inefficient due to a unique character of the event.

Based on the abovementioned reasoning, construct a hierarchical model. The detailed models of food, water, energy, hydrogeology resources would be lower levels of the hierarchical model. In addition, the models already available can be used as well. The less detailed model of social, environmental and economic development will be a higher level of the hierarchical model. The model of higher level defines the corridors of potential opportunities that are the strategies of development. The tactical tasks of development and the specific scenarios belong to the models of lower level.

The interdisciplinary methodology for research of complex systems, called “system dynamics”, is a basis for building the model of social-economic development. A mathematical model of global dynamics has been proposed in (Forrester 1978) and modified in (Gelovani 1974) in the form:

$$\frac{dP}{dt} = P\left(C_B B_M B_P B_F B_Z - C_D D_M D_P D_F D_Z\right), \tag{6.7}$$

$$\frac{dZ}{dt} = PZ_V - Z/T_Z - \left(VU_Z/C_Z\right), \tag{6.8}$$

$$\frac{dV}{dt} = PC_V V_M - V/T_V, \tag{6.9}$$

$$\frac{dR_E}{dt} = -PR_{ME} + \left(VU_{RE}/C_{RE}\right), \tag{6.10}$$

$$\frac{dR_F}{dt} = -PR_{MF} + \left(VU_{RF}/C_{RF}\right), \tag{6.11}$$

$$\frac{dR_W}{dt} = -PR_{MW} + \left(VU_{RW}/C_{RW}\right), \tag{6.12}$$

$$\frac{dS}{dt}=\left[(1+U_S)S_FS_Q-S\right]/T_S, \tag{6.13}$$

$$M_S=C_M(1-S-U_R-U_Z)\frac{V}{P}. \tag{6.14}$$

Here P is the population; V is the capital stock; S is the share of assets in the agricultural sector; Z is the level of environmental pollution; R are the resources; U_R and U_Z are the proportions of capital allocated to resources renewal and antipollution measures, respectively; U_S is the parameter characterizing the capital investment policy in agriculture; M_S is the material standard of living; C_B, C_D, C_M, C_R, C_V, C_Z, T_S, T_V are the model parameters; B_M, B_P, B_F, B_Z, D_M, D_P, D_F, D_Z, R_M, S_F, S_Q, T_Z, V_M, Z_V are functions obtained from analysis of empirical data (they are defined in tabular form).

The various modifications of this methodology were applied to investigate social-environmental-economic development for some countries and regions (Gelovani et al. 2005, Makhov, 2007). In (Sergienko et al. 1997) it was proposed to assess the risk of ecological, economic and social disasters with the help of methodology that combines the dynamic modelling and catastrophe theory. The tabular functions B_F, B_Z, D_P, D_Z, S_F, S_Q were modified to allow for the effect of radioactive contamination on birth rate, mortality, the quality of food, and shutdown of some industrial enterprises and agricultural lands.

To allow for social factors it was assumed that the constants C_B, C_D, C_R, C_V, C_Z, T_S, T_V also as the functions R_M and T_Z are tabular functions of social risk.

6.2.3 *Minimum model of social risk assessment*

The main factors affecting social vulnerability have been were identified with the help of regression analysis. One group of them can be changed by the government through reallocation of resources or adoption of legal measures (control of pensions, stipends, transfer payments, minimum wage, tax policy, etc.). Two factors from this group have been chosen for social risk determination: 1) the ratio between the cost of labor and productivity of labor (Y_1); 2) the share of gross national product (GNP) for the poorest 20% of population (Y_2). The ratio Y_1 in the time interval (t_i, t_{i+1}) was defined as:

$$Y_1=\int_{t_i}^{t_{i+1}} M_s dt \Big/ \int_{t_i}^{t_{i+1}} C_V V_M dt. \tag{6.15}$$

The functions I_k ($k = P, V, R, Z$), that described the sum of flows of variables into the system and into the environment, were introduced in the right-hand side of corresponding equations of model (6.7–6.14).

The effect of the social factor—the share of gross national product (GNP) for the poorest 20% of population (Y_S) on formation and consumption of assess has been introduced by the following assumptions. The rate of asset formation is maximized for some optimal value of income distribution. Passage from this optimal distribution to a more "egalitarian" distribution reduces the incentives to work, because the quality of life becomes less dependent on the results of individual activity. As a consequence, the rate of asset formation decrease. Passage to an income distribution with a greater inequality (reduction of the proportion of income for the 20% of the population) produces a similar effect. Indeed, starting from a certain moment, the intensification of labor will have reached its peak, as determined by physical and technological constraints, and further increase of wages will not produce a corresponding increase in the productivity of labor. To increase the rate of asset formation in this case, we will need to create new jobs to attract an increasing number of workers (including persons from the poorest 20% of the population), who will also require adequate pay for their work. The simplest dependence describing this effect is a quadratic function, although more complex expressions can be used, which describe saturations and the existence of some optimal range of Y_S. In light of the above remarks, the equation for assets V in the model (6.7–6.14) has been modified as follows:

$$\frac{dV}{dt} = PC_V\left(C_{V0} + C_{V1}Y_S - C_{V2}Y_S^2\right)V_M - V/T_V. \tag{6.16}$$

The mathematical statement of the problem has been formulated as follows: given is a system described by equations (6.5–6.14); it is required to solve the Cauchy problem on the time interval [0, T] for a given set of model parameters.

The model has been applied to solve the following problems: 1) prediction of the ecological-economic situation and assessment of the risk of an ecological disaster; 2) calculation of the admissible anthropogenic load Z_{risk} (t) when the risk does not exceed a prespecified value given the actual level of system reserves; 3) calculation of the reserve level R_{risk} (t) required to sustain a given level of risk for the actual anthropogenic load level.

The solution of the Cauchy problem produces the functions $R(t)$, $Z(t)$ and $RISK(t)$ (Problem 1). Then the user-specified risk level and the curve $R(t)$ (Problem 2) or $Z(t)$ (Problem 3) are applied together with expressions (2) and (3) to calculate the functions Z_{risk} (t) or R_{risk} (t).

The model (6.5–6.14) has been used in (Prigogine and Stengers 1984) to solve a number of problems for optimal allocation of funds with the object of minimizing the decrease of material resources. The control variables are U_R and U_Z, which represent the proportion of funds allocated respectively to resource renewal and to antipollution measures; and U_S, which characterizes the policy of investment in agriculture. The optimization problems are solved by the random search method. Single-criterion problems have been considered.

In (Sergienko et al. 1997) a multi-criterion optimal control problem has been considered. This problem maximizes the level of material production and the quality of life, on the one hand, while minimizing the risk and pollution level, on the other.

The problem of optimal risk management has been stated as follows: given is a mathematical model for risk assessment and prediction of consequences of ecological and technogenic disasters (6.1), subject to a system of constraints that define the admissible range of model variables and controls; it is required to find the controls control $U_R(t)$, $U_S(t)$ and $U_Z(t)$ that minimize the functional:

$$\begin{aligned} F(U_R, U_S, U_Z) = {} & \frac{A_1 C_{P_1}}{I_P} - A_2 C_{P_2} m_P + A_3 C_Z I_Z + A_4\left(\frac{\max Z}{Z_0} - 1\right) \\ & + A_5\left(\frac{Z}{Z_0} - 1\right) + \frac{A_6 C_R}{I_R} + A_7 C_{risk} I_{risk} + \frac{A_8 C_{MS1}}{I_{MS}} - A_9 C_{MS2} m_{MS} \\ & + A_{10}\left(1 - \frac{Q_l}{Q_{l_0}}\right) + A_{11}\left(1 - \frac{F_R}{F_{R_0}}\right) - A_{12} C_{F_R} m_{F_R}. \end{aligned} \tag{6.17}$$

Here $A_i, i = \overline{1,12}$, are the weights; $C_{P_1}, C_{P_2}, C_Z, C_R, C_{risk}, C_{M_{S1}}, C_{M_{S2}}, C_{F_R}$ are multipliers the contribution of each term in the objective function; m_P, m_{M_S}, m_{F_R} – are the minimum values of the derivatives of the corresponding variables on the interval $[t_0, t_1]$; Z_0, Q_{l_0}, F_{R_0} are the initial values at time t_0;

$$I_P = \int_{t_0}^{t_k} P(t)\,dt; \tag{6.18}$$

$$I_Z = \int_{t_0}^{t_k} Z(t)\,dt; \tag{6.19}$$

$$I_R = \int_{t_0}^{t_k} R_E(t)\,dt; \tag{6.20}$$

$$I_{MS} = \int_{t_0}^{t_k} M_S(t)\,dt; \tag{6.21}$$

$$I_{risk} = \int_{t_0}^{t_k} RISK(t)\,dt. \tag{6.22}$$

max Z is the maximum pollution level on the interval $[t_0, t_1]$; Z, Q_l, F_R are the values at time t_k.

$$Q_l = Q_l(M_S, F_F, Z, P), \tag{6.23}$$

$$F_R = F_R(P, Z, S), \tag{6.24}$$

where Q_l and F_R are tabular functions that respectively describe the quality of life and the nutrition level.

The optimization problem has been solved by the random search method. Solution of optimization problem minimizing the quality of life has produced the optimal control dynamics that minimizes the pollution level subject to maximum attainable quality of life.

The investigation of social, economic and environmental sustainable development to determine the scenarios of efficient food, energy and water use based on above mentioned methodology has been carried out in (Atoyev et al. 2014). A flowchart of the interlationships in the modified model (Sergienko et al. 1997) is shown in Fig. 6.4.

The following additional indicators were used as the model variables: food (R_F), water (R_W) and energy (R_E) resources; capital shares to provide material standards of living (U_L) and restore food (U_{RF}), water (U_{RW}) and energy (U_{RE}) resources. The model also takes into account the parameters characterizing measure of vulnerability of society due to increasing uncertainty related with global changes (including climatic changes), migration of population and transboundary flows of capitals, resources and pollutions.

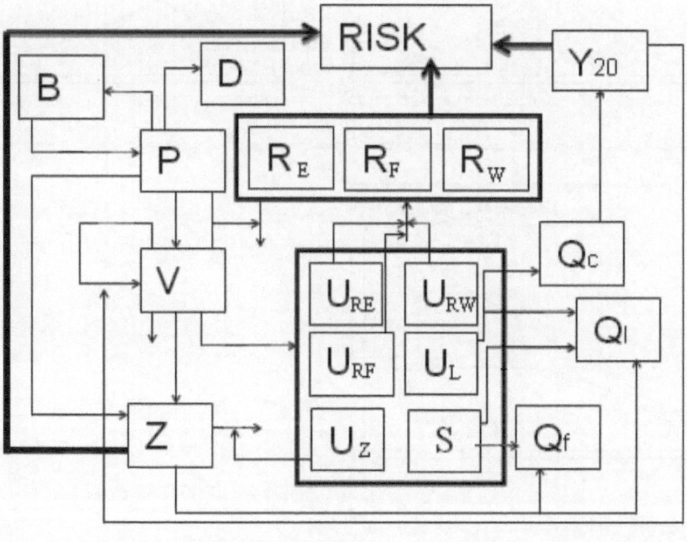

Figure 6.4. Block scheme of model for investigation of risk impact on food and economic fields.

The model parameters are determined on the basis of empirical data. Some of these parameters are viewed as functions of social risk dependent on the probability of sociogenic catastrophes (Moiseev et al. 1985). Computing the social risk is based on the following data: 1) labour cost-productivity ratio; 2) differentiation level for total incomes of decile population groups; 3) efficiency of state governance based on the GRICS (Governance Research Indicator Country Snapshot) index.

The results of modelling for one of region of Ukraine (Chernobyl zone) are presented in Fig. 6.5.

The circles denote the real data corresponding to the 1991–2010 samples for the Chernobyl zone. The curve 1 marks the solution for dynamic problem under initial values of parameters. The curve 2 marks the solution for identification task.

The results of solving optimal control are presented in Fig. 6.6. The model parameters have been determined on the basis of real data analysis as a result of solving the identification tasks over the interval of years 1991–2010. The curve 1 corresponds to the dynamic problem, and the curve 2 – to the optimal control problem.

The ranking of Ukrainian regions by the levels of socio-economic and ecological-technogenic threats has been carried out on the basis of results obtained for all regions of Ukraine (Fig. 6.7). The highest levels of ecological threats, as demonstrate obtained results, are in Donetsk, Dnipropetrovsk and Luhansk regions, the smallest—in Chernivtsy, Lviv and Zakarpattya regions. The highest levels of economic threats are in Kirovograd, Zakarpattya ant Kyiv regions, the smallest—in Chernigiv, Volyn, Sumy, Rivne and Vinnitsia regions.

The highest level of technogenic threats are in Odesa, Volyn, Donetsk, Zaporizhya and Lviv regions, the smallest—in Rivne, Zhitomir, Khmelnitsky, Sumy and Vinnitsia regions. The highest level of social threats is in Luhansk, Donetsk, Volyn and Kirovograd regions, the smallest in Zakarpattya, Ivano-Frankivsk regions and Kyiv (Atoyev 1999).

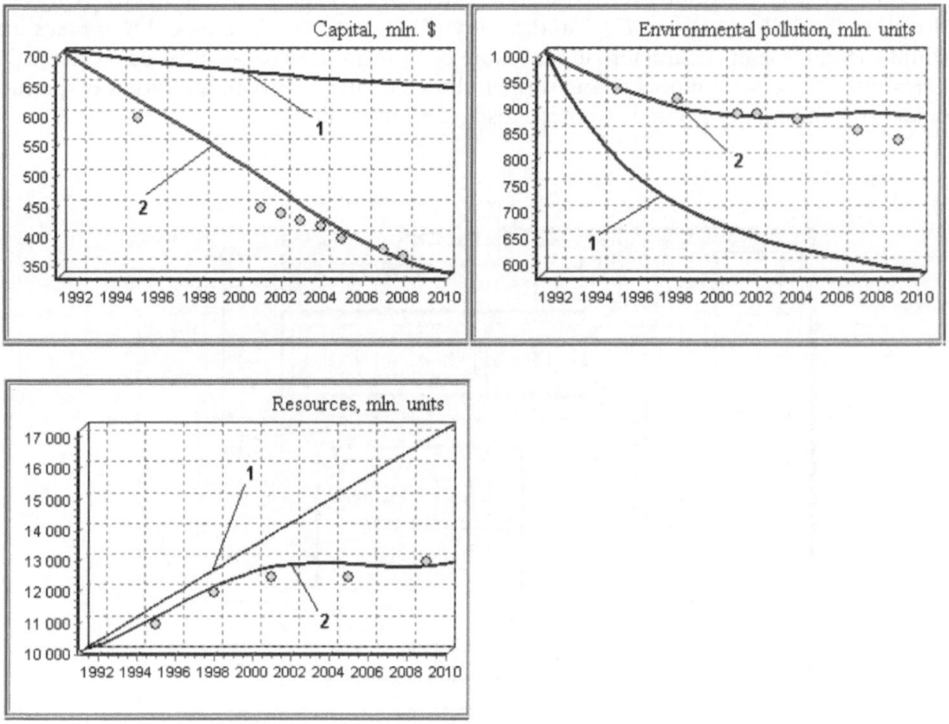

Figure 6.5. The results of identification task solution.

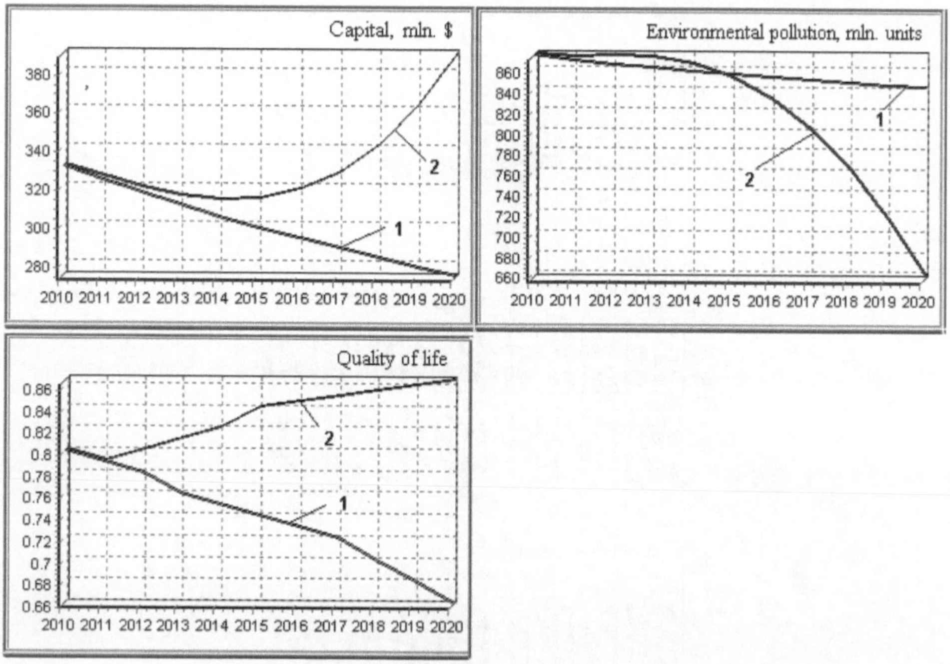

Figure 6.6. The results of optimal control problem.

The profile of threats was calculated for each region. The results for Mykolayv region are presented at Fig. 6.8. Mykolayv is classified as region of high threat level. The index of economic threat $A_1 = 0.55$ (22-d place).

The index of social threat $A_2 = 0.69$ (8-th place). The index of technogenic threat $A_3 = 0.46$ (8-th place). The index of ecological threat $A_4 = 0.19$ (12-th place). The integral index of removing to low threat level is equal to 0.437661 (3-d place among 9 regions of group with high level of threat). The ecological-technogenic component of this index is equal to 0.244434 (3-d place). The social economic component is equal to 0.348658 (3-d place).

6.2.4 *The decision support system for environmental security assessment*

The decision support system for integrated security assessment and management has been elaborated. The distinctive feature of elaborated computer system is joining in technological gear software that allows the forecasting and optimization tasks solution that minimize the level of environmental risk. It allows calculating the optimal controls, minimizing the negative impacts of Global Changes on sustainable development. The impacts of man-made and natural disasters are investigated with the help of various mathematical models. The control board of computer system is presented in Fig. 6.9.

The ranking of 25 regions of Ukraine on a degree of environmental vulnerability (EV) and its technogenic, ecological, social and economic components was carried out. The projections of catastrophes on the surfaces of technogenic and ecological security are shown in the middle of Fig. 6.9.

The changing phase portraits that correspond to the dynamics of chosen environmental impacts, also as changing operating life at chosen load are presented at left part of Fig. 6.9.

The ranking of regions on EV and its components that were calculated with the help of examined method are presented at the bottom. Solid vertical for components corresponds to medial value of threat indices for Ukraine. Solid vertical on the figure for integral threats

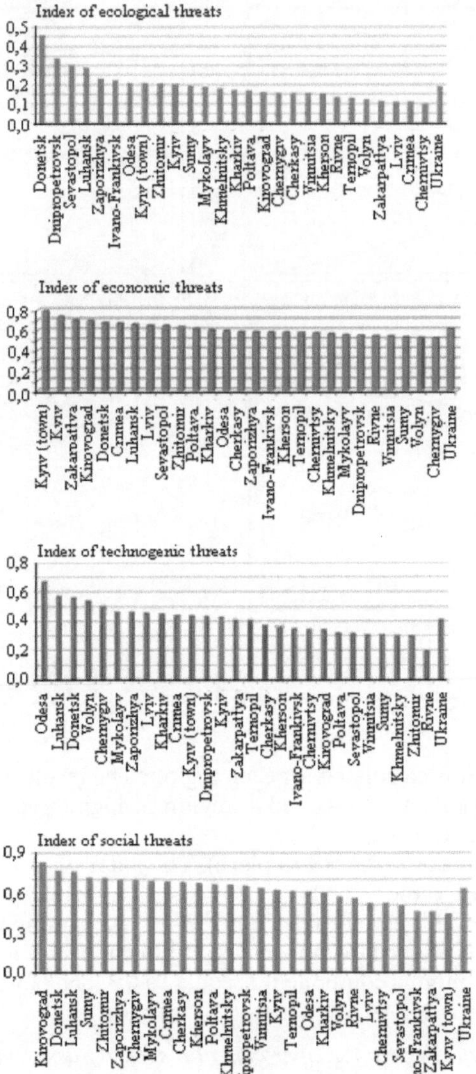

Figure 6.7. The ranking of Ukrainian regions on levels of threats.

separates regions with low and middle level of threat, dashed vertical separates regions with middle and high EV.

6.2.5 *Calculation of the optimal controls, minimizing the decreasing of complex system's stability*

Let $\boldsymbol{A}_N$ be a vector of system parameters, characterizing its normal functioning, $\boldsymbol{A}_C$ – a vector, characterizing system in catastrophic state, $\boldsymbol{A}_i$ – vector of current state of system. All state indices are forming as follows:

$$\Delta_C = \sqrt{\sum_{i=1}^{S} (A_{iN} - A_{iC})^2}, \tag{6.25}$$

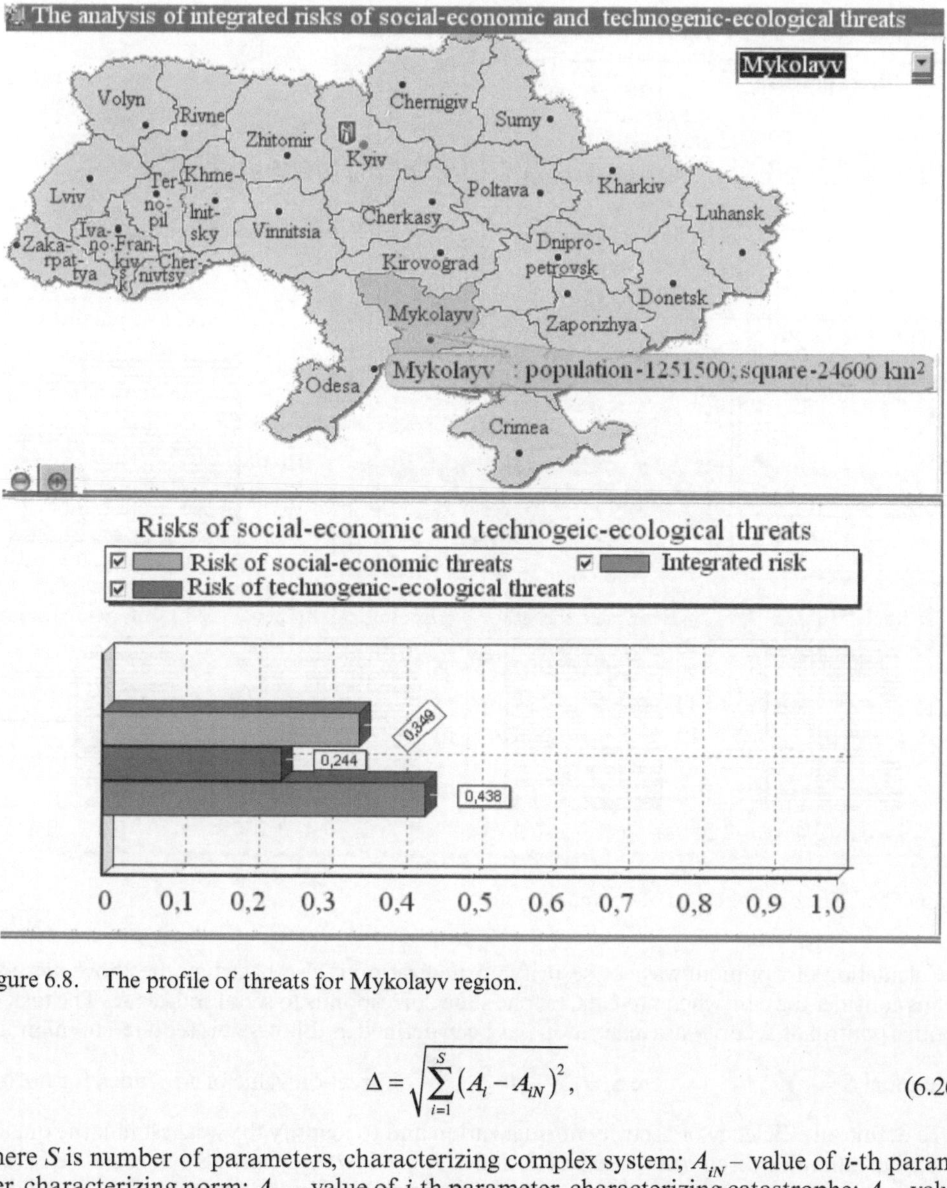

Figure 6.8. The profile of threats for Mykolayv region.

$$\Delta = \sqrt{\sum_{i=1}^{S} (A_i - A_{iN})^2}, \tag{6.26}$$

where S is number of parameters, characterizing complex system; A_{iN} – value of i-th parameter, characterizing norm; A_{iC} – value of i-th parameter, characterizing catastrophe; A_i – value of current i-th parameter; Δ_C – index of complex system state corresponding to catastrophe; Δ – index of complex system current state. If $\Delta > \Delta_C$, then we consider $\Delta = \Delta_C$.

Risk of transfer from one functional state into another (R) is determined as follows: $R = \Delta / \Delta_C$. If $\Delta_C = 0$, then a transfer from one state into another is considered to be impossible. The distances between current state and states that characterize the transfers into pre-catastrophic and catastrophic states (if such transfer is possible) can serve as valuation of reserve. Total risks of transfer into pre-catastrophic and catastrophic states are determined as follows:

$$R_{\Sigma j} = \left(\sum_{i=1}^{4} R_{ij}^2 \right)^{1/2}, \tag{6.27}$$

where $j = 1$ corresponds to pre-catastrophe, $j = 2$ – to catastrophe. If $R_\Sigma > 1$, it means that $R_\Sigma = 1$.

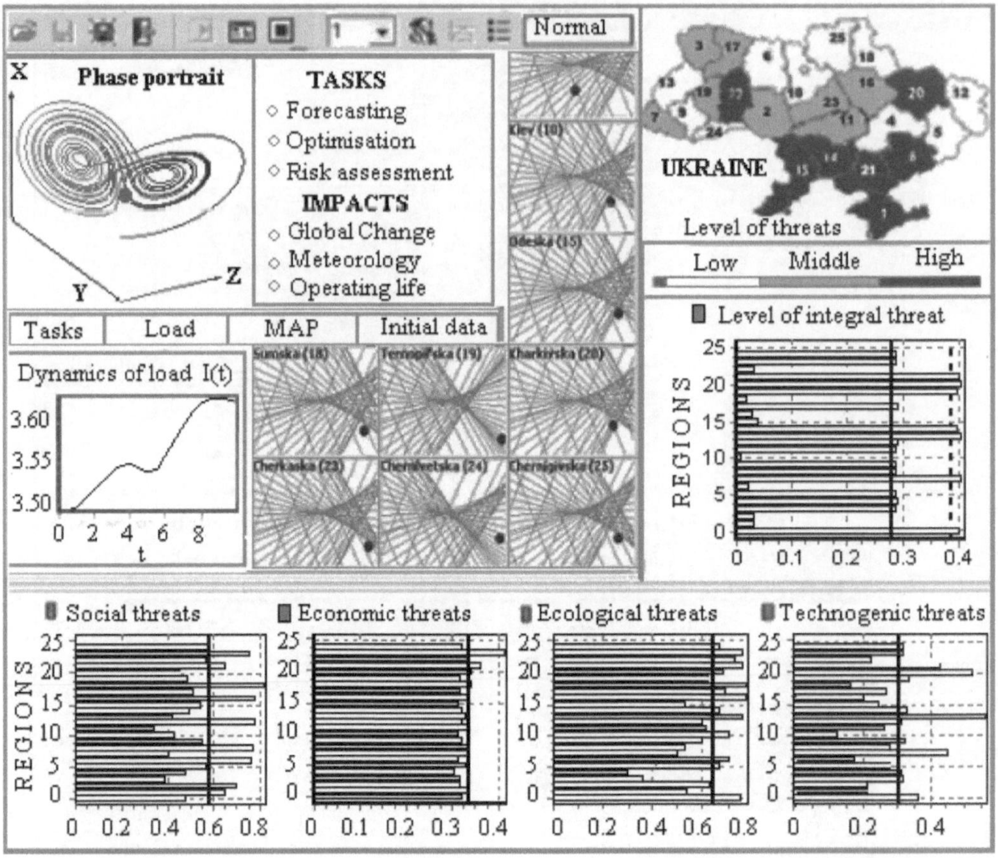

Figure 6.9. The control board of computer system.

Calculation of optimum ways of security normalization is also based on the above method. Let us consider the case when pre-catastrophe state corresponds to set of indices A_i. The task of optimal control of security normalization has been defined as follows: it is required to minimize functional $S = \left(\sum_{i=1}^{4} S_i^2 \right)^{1/2}$, where $S_i = |A_i - A_i^N|$, A_i^N – bifurcation value of i-th index for norm.

To define an efficiency of security normalization and to identify the weakest link the quality of optimization has been introduced as an index $I_{iopt} = S^i_{opt}/S^i_{Cat}$, where S^i_{Cat} – distance from a current state of *i*-th index up to bifurcation value (norm), S^i_{opt} – the same distance found during solution of optimization task. Besides the values describing distance from a current state up to bifurcation surface before and after optimization are entered, accordingly S_{Cat}, S_{opt}.

To solve optimization problem a modified random search method—statistical gradient method—was used. The choice of the method was caused by lack of analytical expression for objective function and by derivatives' calculation complexity.

6.3 COMPLEX BIOLOGICAL SYSTEMS

An organism in contrast with a machine is a self-reproducing, self-organizing whole. According to (Capra 1996), the term "self-organization" as a definition of the nature of live organisms has been proposed by Immanuel Kant. The main contradiction of the current medicine appears to be the fact that it attempts to treat not a patient but some organism subsystems, whereas an illness is a consequence of structural changes on molecular and subcellular levels. It is getting developed based on various algorithms, simultaneously affecting different subsystems and organs.

Even in those cases where a performed therapy provides with a positive effect, there is no guarantee that it is not achieved due to reduced functionality of some other organs. The side effects of such therapy can intensify de-synchronization of various regulatory mechanisms, break a balance of various controlling systems, and cause a vicious circle of pathological disturbances.

It will be possible to move from the medicine of "consequences" to the medicine of "reasons" and to work out adequate tactics of patient's treatment, only if we take into account individual features of an organism as a whole.

The direction of the therapy should become:

1. a support and stimulation of the organism protective functions, upraising its adaptiveness and balancing individual controlling systems and their communication;
2. an achievement of maximum positive effect of used medicines with a minimum side effect on other subsystems of the organism;
3. a development of a combined therapy optimal dynamics to ensure positive changes in reserve abilities of the organism as a whole.

Mathematical modeling becomes an important auxiliary tool to solve these problems. It upraises considerably efficiency of system analysis of contradictory experimental data, transforming it into new understanding, based on integrated parameters of environmental and radiation effect on an organism as a whole and its individual systems. Better approximation of radiation and other environmental factors impact on the organism can be achieved using such an approach.

Computer technologies are to be used as learning tools to teach a physician to treat an organism as a whole in opposite to a specific illness, to see an integrated picture of interconnected organism's systems, to predict consequences of stress impact, to individualize the treatment process (choice of an optimal treatment strategy).

In (Atoyev 2005) advanced modeling method for health risk assessment has been elaborated. It is based on catastrophes, bifurcation, and chaos theories. The 4-dimensional typology of the organism network as a self-reproducing, self-organizing entity, has been proposed. A nonlinear mathematical model of neuro-immune-endocrine network (NIEN) has been elaborated. It makes possible to investigate interrelations between various control mechanisms of metabolic and hormonal regulation, radiation, and stress impacts on thyroid system, mechanisms of various endocrine diseases formation.

The following tasks have been examined:

1. comparative analysis of endocrine system regulation for some types of thyroid system pathology (such as hypo- and hyperthyroidism, T3- and T4-thyreotoxicoses);
2. optimum control of thyroid system restoration;
3. ranking of radiation influence on various links of thyroid system.

The performed system data analysis made it possible to reveal the most probable sequence of metabolic and hormonal changes leading to a thyroid pathology occurrence.

> "We are but whirlpools in a river of ever-flowing water.
> We are not stuff that abides, but patterns that perpetuate themselves."
> Norbert Wiener, The Human Use of Human Beings, p.96 (1950).

6.3.1 *The mathematical model of health risk assessment*

A human organism (HO) can be considered as an open, non-Hamiltonian, non-linear, dissipative, potential, self-reproducing, self-organizing system with networking elements, near an unstable steady state far from equilibrium, in bifurcation zone where any fluctuation leads to an unpredictable future (Prigogine and Stengers 1984). A disturbance in some network node may lead to catastrophic consequences in the whole network. The effects of a disturbance could impact areas far from the initial source due to existing longer-term links and a great extent of cause and effect in HO. Health risks are those insidious risks emerging slowly but surely over time. Any environmental impact increases possibility of the case where some fluctuation can cascade into a disease with unpredictable health consequences.

The high initial conditions susceptibility is a common feature of such systems and it makes it impossible to predict patient's health or a treatment effectiveness. Traditional methods to predict a system development are not enough efficient. There is a "growing recognition that risk management should not be based on the assumption of linear cause and effect", as noted in (Wilkinson et al. 2003, p. 387). Such entities are usually examined using recent theories of nonlinear systems, synergetic, chaos, and bifurcation.

Let us propose that organism satisfies all the requirements of potential system and can be described by some potential function $U(X)$ of variable X. This function has some steady states; some of them are stable, others—unstable. The transformations of system from one steady state to another, or transformation of a character of a steady state (for instance, from stable to unstable) are functions of control parameters $\boldsymbol{A}$. These parameters control both a movement of system's current position upon surface U, and transformations of the surface itself.

Let X be a level of health. To make it simple, it is possible to put forward following suppositions. Potential function $U(X,\boldsymbol{A})$ has three stable steady states (Fig. 6.10). The first one (X_1) characterizes normal condition. The level of health in this state is maximal. The second one (X_2) characterizes premorbid or pre-pathological state with a medium level of health (the reserves of the organism and its adaptiveness are lowering, but yet not across the critical level which separates normal functioning and an illness. The third one (X_3) characterizes disease.

The risk estimation of sudden and unexpected transformation from various states of organism has been performed using the algorithm based on theory of catastrophes (see part 6.3).

So, the set of values of the potential function $U(X,\boldsymbol{A})$ control parameters and their deviations from bifurcation surfaces may be utilized for risk matrix determination to calculate probabilities of transformations between various steady-states of physiological system. Thus, the potential function $U(X,\boldsymbol{A})$ determines the level of health. One of the catastrophes' theory universal deformations can be used to define $U(X,\boldsymbol{A})$ depending on the number of control parameters.

Let us introduce some postulates that we use as a basis to develop the model.

1. A state of an organism is considered as complex interrelationship in a "square of health" (SH), which is determined by a state of NIEN regulation, cellular, and physiological adaptiveness, mental sphere, and a genomic pre-history. A set of health threats, including environmental and radiation impact, has a potential to destroy an intricate balance existing within the SH by altering the above-mentioned factors. The balance disturbance within the SH is a most formidable health threat to be prevented, as it leads to a transition from norm to diseases.

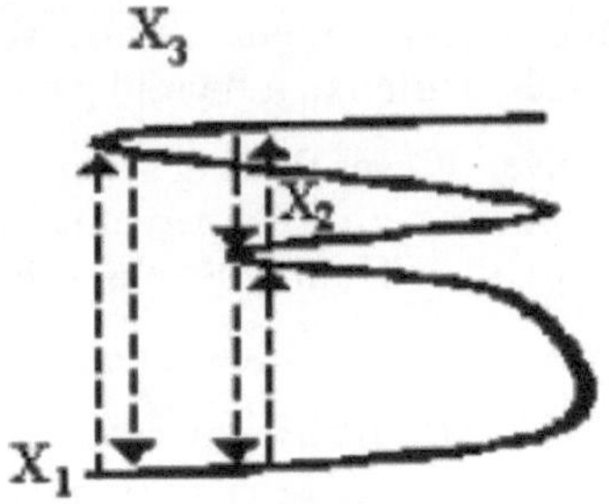

Figure 6.10. Transformation between various steady states.

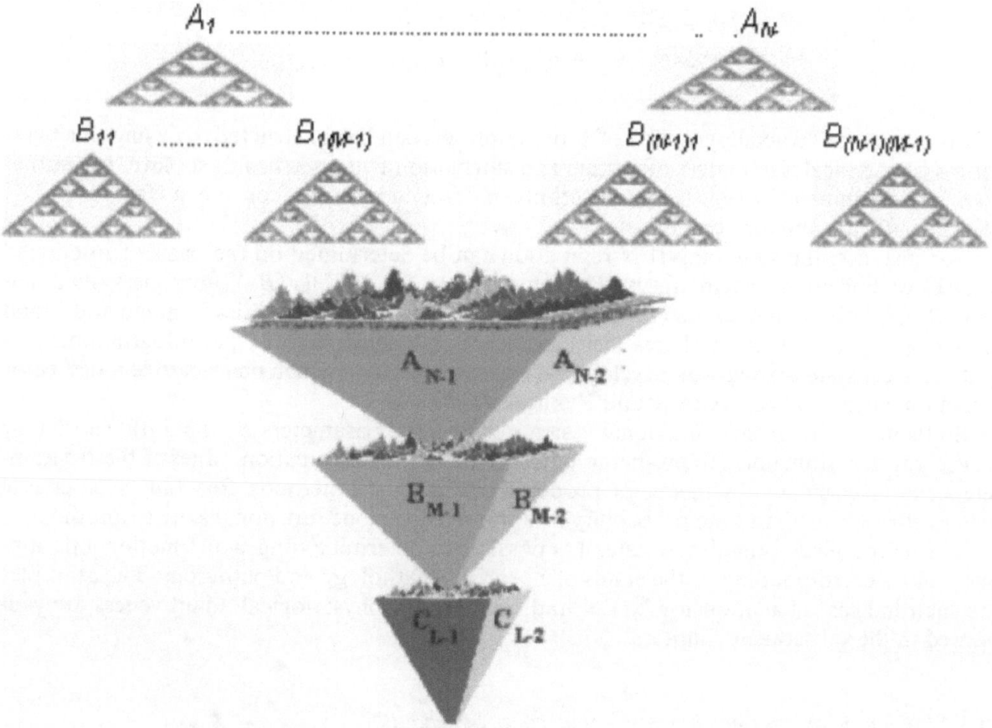

Figure 6.11. Fractal pyramid of bifurcation sets.

2. To examine a health environmental impact problem we have chosen a 4-dimensional typology. The transformations of health are defined by one of the universal deformations of catastrophes' theory—a "butterfly":

$$-\partial U(X,A)/\partial X = X^5 + A_1X^3 + A_2X^2 + A_3X + Aa_4, \tag{6.28}$$

where A_i (i = 1,4) are control parameters, characterizing various aspects of the organism's functioning (i = 1 – index of NIEN regulation, determined by hormones levels, energetic substrates, sub-populations of lymphocytes, etc.; i = 2 – index of cellular and physiological adaptiveness to environmental impacts, determined by the balance between energy synthesis and consumption processes, reserve possibilities to extreme loads, etc.; i = 3 – index of mental health, determined by psychical and psychological peculiarities of the organism; i = 4 – index of genomic pre-history, determined by genetic inclinations of the organism).

According to Prigogine's theory, dissipative structures can not only maintain themselves in a stable state far from equilibrium but may even evolve. When the flow of energy and matter through them increases, they may go through new instabilities and transform themselves into new structures with increased complexity. That is why organism (as pointed by F. Capra), can be considered as a giant web of life, which consists of fractal networks where any network contains another one. And a node of the network being zoomed is getting transformed into a new network (Capra 1996).

Under more detailed investigation, parameters Ai also may be examined as n-dimensional functions $A_i(Fj)$ (j = 1,n), that are arguments of some potential functions U_{Ai}. Variables Fj describe each index in more detail. Such iterative process leads to the next sublevel of bifurcation sets, determined by:

$$
\begin{aligned}
&dA_i / dt = -\partial U_{A_i} / \partial A_i, \\
&A_i^M + B_{i1}A_i^{M-2} + \cdots + B_{i(M-3)}A_i^2 + B_{i(M-2)}A_i + B_{i(M-1)} = 0.
\end{aligned}
\tag{6.29}
$$

So, a peculiar "fractal pyramid" of bifurcation sets can be constructed; by using this pyramid a hierarchical risk matrix of system transformation from one health sublevel to another can be determined. In Fig. 6.11 projections of "fractal pyramid" on the surfaces $A_{N-1}A_{N-2}$, $B_{M-1}B_{M-2}$, $C_{L-1}C_{L-2}$ are presented.

For instance, the state of NIEN regulation can be determined on the basis of integrated indices of hormone system status (B_{11}), energetic status of cells (B_{12}), immune system status (B_{13}), blood system status (B_{14}), characterizing endocrine, energetic, immune and blood systems, respectively. A mathematical method of risk assessment of pathological and pre-pathological state arising was developed using these indices, which characterize a degree of functional disturbances (Atoyev and Rykhtovsky 2000).

Shifts in the organism's functional systems change the parameters B_{1j} ($i = 1,4$). Those may cause transfers from norm to pre-pathology or to pathology. Bifurcation values of these parameters can be calculated by means of proposed mathematical methods. Reaching such critical values abruptly increases the probability of a transfer from one functional state to another.

Thus, for a given organism's state, it is possible to determine ranges of functional parameter values corresponding to the states of norm, pre-pathology and pathology. The examples of such indices calculation for NIEN and cellular and physiological adaptiveness are presented in the subsequent chapters.

6.3.2 *Chaotic oscillations in a system of energy exchange*

The analysis of an energy exchange shows that while such dissipative structures as an organism receive their energy from outside, the instabilities and jumps to such new forms of organization as a disease appear to be a result of fluctuations amplified by positive feedback loops (Capra 1996, Prigogine and Stengers 1984).

Let us examine these feedback loops more deeply to investigate possible mechanisms of chaotic behavior developing in the energy conversion system. Based on experimental investigation of cellular energy exchange (Thabrew et al. 2004), the following postulates can be proposed:

1. Oxidative phosphorylation in mitochondria plays a central role in the energy supply of various cellular functions: active ion transport, proliferation, muscle contraction, protein synthesis, etc.
2. An organism has some demand on activities of various physiological systems, cellular, and subcellular networks.
3. Some part of the cell energy is getting consumed for internal cellular functions; another part is used for external functions, determined by the organism demand.
4. The demand depends on various factors: current states of physiological systems, loads, environmental impacts, hormonal pool, efficacy of control mechanisms, etc. Let $b_{1i}(t)$ denote a current demand on activity of i-th control system normalized to the energy synthesis unit (mitochondrion) – Y_i.
5. The current state of i-th control system is characterized by a level of its external function which determines its supply to the organism demand. The supply may correspond to the demand or may not. Let $b_{2i}(t)$ denote a level of supply normalized to the unit of i-th control system's function (a contractility for muscle, a rate filtration for a kidney, etc) X_i. According to the proposed designations, $b_{1i}(t)Y_i$ and $b_{2i}(t)X_i$ characterize total demand on activity of i-th system and total supply on fulfillment of external function determined by possibilities of mitochondria, correspondingly.
6. Let the rate of supply change be proportional to the prevalence of organism demand over supply. Then we have the following equation for a dynamic of external function:

$$dX_i / dt = a_{1i}(t)\left[b_{1i}(t)Y_i - b_{2i}(t)X_i\right], \tag{6.30}$$

where $a_{1i}(t)$ is a parameter characterizing adaptive possibilities.

7. Some cellular mechanisms exist that modify intracellular structures and involve new mitochondria in energy synthesis under prevalence of demand (biosynthesis, redistribution of ion and energy currents, etc.). Let us assume that the higher X_i is the greater is a number of mitochondria Y_i involved in the process. Let $c_{1i}(t)$ be a demand on an increasing of number of mitochondria normalized to X_i unit, and $c_{2i}(t)$ be a share of Y_i involved in X_i maintenance. Thus, $c_{1i}(t)X_i$ and $c_{12i}(t)Y_i$ characterize total demand and supply on mitochondria correspondingly.
8. Let the rate of mitochondria change be proportional to a prevalence of energy demand over supply. Then we have the following equation for a dynamic of mitochondria involved in energy synthesis:

$$dY_i / dt = a_{2i}(t)\left[c_{1i}(t)X_i - c_{2i}(t)Y_i\right], \tag{6.31}$$

where $a_{2i}(t)$ is similar to $a_{1i}(t)$.

9. The fact that an increase of Y_i is limited by various biological factors (for instance, by a competition for substrates) is not taken into account in (6.10). Moreover, some part of Y_i is involved in cellular internal functions. This part is proportional to X_i (which adds to an existing demand to *i*-th system structures). It is also proportional to the level of pathological disturbances Z_i developing in cellular structures and demanding additional energy for their elimination. Consequently, the equation (6.31) is to be transformed into:

$$dY_i / dt = a_{2i}(t)\left[c_{1i}(t)X_i - c_{2i}(t)Y_i - c_{3i}(t)X_iZ_i\right], \tag{6.32}$$

where $c_{3i}(t)$ is a demand on Y_i caused by a maintenance of cellular internal functions and normalized to units of external function and disturbances that are being developed in the cell.

10. The rise of disturbances is proportional to the load level on the structures maintaining external function X_i. It depends on X_i, and the number of Y_i involved in the energy synthesis. We also propose the existence of some control (adaptive) mechanisms, which eliminate some part of disturbances under some restrictions. These mechanisms are getting activated under appearance and accumulation of disturbances. Let $d_{1i}(t)$ and $d_{2i}(t)$ be the constants of proportionality. Thus, the rate of disturbances will be described by the equation:

$$dZ_i / dt = \left(d_{1i}(t)X_iY_i - d_{2i}(t)Z_i\right). \tag{6.33}$$

Let us examine one isolated cell that has one external function. Let all model coefficients be constant. Introduce the following designations:

$$\begin{aligned} &t = t^* / (a_2c_2), \sigma = (a_1b_2)/(a_2c_2), r = (b_1c_1)/(b_2c_2) \\ &b = d_2 / (a_2c_2), X = d_1X(c_3/d_1)^{1/2} / (a_2c_2) \\ &y = d_1b_1Y(c_3/d_1)^{1/2} / (b_2a_2c_2), z = (b_1c_3Z)/(b_2a_2c_2). \end{aligned} \tag{6.34}$$

In this case equations (6.31), (6.33) and (6.34) become:

$$dx / dt^* = \sigma(y - x), \tag{6.35}$$

$$dy / dt^* = rx - y - xz, \tag{6.36}$$

$$dz / dt^* = xy - bz. \tag{6.37}$$

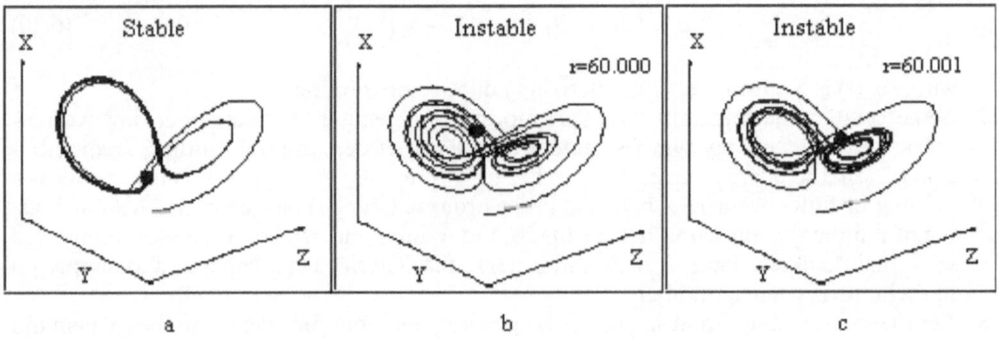

Figure 6.12. Impact of fluctuation on dynamics of Lorenz attractor.

According to Lorenz (1963), sustained chaotic behavior arises in this model. The rise of demands leads to increasing of parameter r. The rise of supply leads to its decreasing. There are intervals of demand/supply ratio corresponding to different regimes of metastable chaos.

Small fluctuations of parameter r case the essential transformation of system dynamics. At the upper boundary of above mentioned intervals of demand/supply ratio such fluctuations may lead to catastrophic consequences connected with transition from stability to instability, from a state with a low level of cellular vulnerability to a state with high one.

A transition, from stability (low level of vulnerability) to instability (high level of vulnerability) could be analogous to the transition from a laminar to turbulent flow. The modeling of such transition presented in Fig. 6.12 illustrates the impact of fluctuation (parameter r) on dynamics of phase portraits. Fig. 6.12a corresponds to stable or laminar dynamics ($\sigma = 4$, $r = 80$). Fig. 6.12b and Fig. 6.12c—to instable or turbulent dynamics). Small fluctuation of parameter r (from $r = 60.000$ to $r = 60.001$) cases the essential transformation of phase portrait.

Thus, relationships between cellular external function, cellular energy, and cellular disturbances may be described by the Lorenz model of metastable chaos. A transition, from stability to instability of energetic flows, could be analogous to the transition from a laminar to turbulent flow.

6.3.3 *Mathematical model of neuro-immune-endocrine regulation*

Analysis of experimental data concerning various aspects of radiation negative influence on human organism unambiguously testifies that one of its most appreciable consequences is a harsh spike of the diseases dealing with Thyroid System (TS) function disturbances. These sorts of diseases are progressing during all after-Chernobyl period.

Despite the intensive research carried out lately, there is no uniform vision of the exact TS links and regulatory mechanisms which are the main targets for radioactive iodine. What are the chains of events which are getting realized with the greatest probability under a radiation action? Which links are primary targets of radiation action, and which ones are just of a secondary, or compensatory character? Without the answers to these questions it is impossible to develop an effective prevention under radiation action on a human organism.

So, detailed analysis of TS control mechanisms appears to be one of the most actual problems in modern endocrinology. A weakening of the TS function triggers a consequent sequence of hormone and metabolic violations, resulting in the development of various pathological processes. Thus, it is also essential to determine an effective therapy strategy which would lead to a synchronization of various regulatory mechanisms activities and optimization of the usage of specific medicines, normalizing levels of Thyroid Hormones (TH) during different types of TS pathology.

Cyclic Nucleotides (CN), cell membrane system (the activity of calcium and sodium pumps), Corticotrophin-Releasing Factor (CRF), corticotrophin (ACTH), Thyrothropin-Thyrostimulating Hormone (TTH), Thyrotropin-Releasing-Factor (TRF), Triiodothyronine (T3), thyroxin (T4), interleukin-1, Arachidonic Acid (AA), Catecholamines (CC), Glucocorticoids (GC), and a number of other hormones and metabolic factors play an important role in the control of NIEN (Sobolev 1987).

Interleukin (IL-1) appears to be one of the most important links between immune and endocrine system. It is being produced not only by cells participating in myelo- and lymphopoesis (monocytes, macrophages, neutrophyle granulocytes, B-lymphocytes, natural killers, and so on), but also by the cells of central and peripheral nervous systems (brain neurons, peripheral neurons and others). IL-1 activates hypothalamus-hypophysis-adrenocortical system; it stimulates synthesis of CRF and somatostatin, causes alterations in noradrenergic neurons' activity. A high level of IL-1 leads to GC and ACTH levels growth in blood and inhibits TTH synthesis. This inhibition consequently causes drop of TTH, T3 and T4 levels. GC also causes an inhibition of IL-1 synthesis.

Block-scheme of radiation impact on NIEN is presented at Fig. 6.13. Shvitra's TS model based on non-linear equations with a delay (Shvitra 1989), has been modified introducing some terms to take into account impacts of metabolic and stress factors (including radiation) on NIEN (Atoyev 2005).

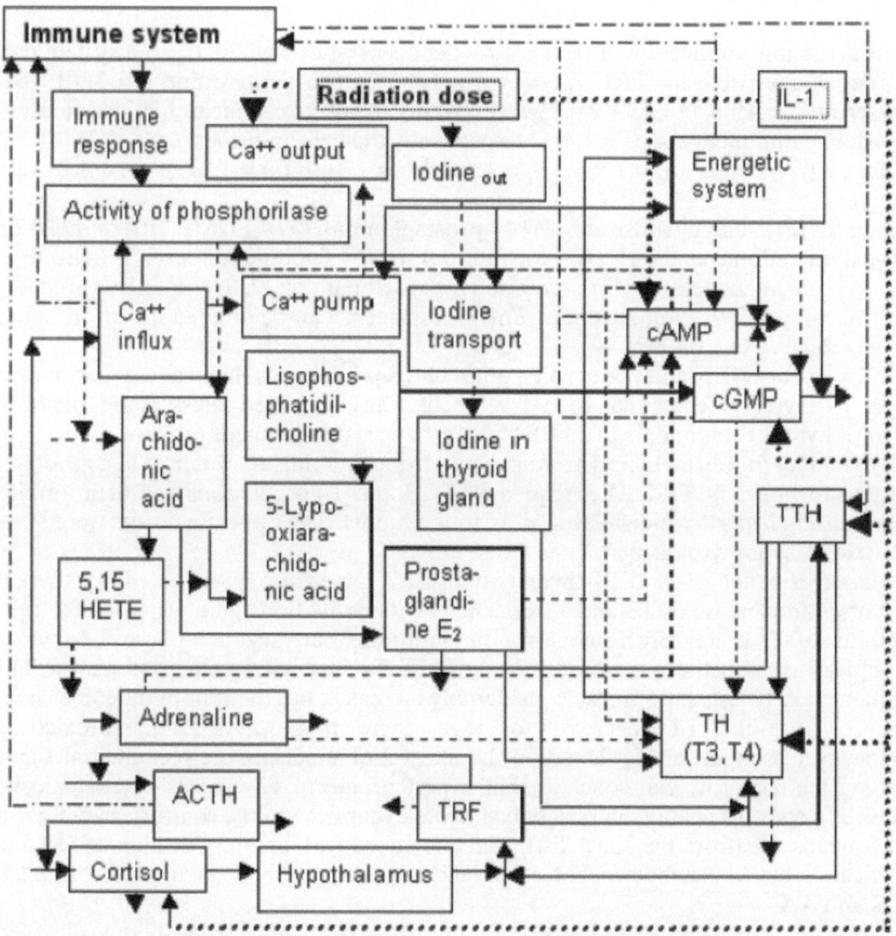

Figure 6.13. Block-scheme of NIEN. Solid and dashed lines are used for hormonal control of TS, dash-and-dot line—for interrelation between immune and endocrine systems, and dotted line—for radiation impact on NIEN.

$$
\begin{aligned}
\frac{dT_1}{dt} &= a_1\{\beta_1 + \beta_2 Ah(t)[\delta_1 + IL1(t)] + \delta_2 IL1(t) - \delta_3 S(t)\} \\
&\quad \{1 + a_1[1 - T_3(t)/K_3] + a_2[1 - T_4(t)/K_4]a - d_1 T_1(t-h_1)/K_1\}T_1(t), \\
\frac{dT_2}{dt} &= a_2\{[\gamma_1 + \gamma_2 A(t) + a_2 + a_3 - \delta_4 GH - \delta_5 D_N(t) - \delta_6 IL1(t) - \delta_7 C(t) + \delta_8 E_s(t)], \\
&\quad d_2 T_1(t)/K_1 - a_3 T_3(t)/K_3 - a_2 T_4(t)/K_4 - d_3 T_2(t-h_2)/K_2\}T_2(t)\}T_2(t), \\
\frac{dT_3}{dt} &= a_3\{\beta_3[\gamma_3 + \gamma_4 v_1 + \gamma_5 z \Pr(t)]d_4 T_2(t)/K_2 \\
&\quad + [1 - \beta_3]T_4(t)/K_4 - d_5 T_3(t-h_3)/K_3\}\}T_3(t), \\
\frac{dT_4}{dt} &= \alpha_6\{[\gamma_6 + \gamma_7 v_2 + \gamma_8 \Pr(t)]d_6 T_2(t)/K_2 - d_7 T_4(t-h_4)/K_4\}T_4(t), \\
\frac{dCa}{dt} &= c_1\{[c_2 + c_3 T_2(t)][1 - \exp(-\alpha_5 t)]\exp(-\alpha_6 t) - P_{Ca}(t)\}T_4(t),
\end{aligned}
\tag{6.38}
$$

$$
\begin{aligned}
v_1 &= [k_1 + G(t)][k_2 - A(t)][k_3 + LP(t)][k_4 + I(t)]E(t)/[b_2 + E(t)], \\
v_2 &= [k_5 + G(t)][k_6 - A(t)][k_7 + LP(t)][k_8 + I(t)]E(t)/[b_3 + E(t)],
\end{aligned}
\tag{6.39}
$$

where $Ca(t)$ – ions of calcium, $T_i(t)$ ($i = 1.4$) – concentrations of the thyroid system components, namely: respectively, TRF – ($i = 1$), TTH ($i = 2$), triiodothyronine, ($i = 3$), thyroxin in blood serum ($i = 4$); h_i ($i = 1.4$) – delays describing, respectively, mean lifetime of each thyroid system components; K_i ($i = 1.4$) – respectively, their mean values; a_i($i = 1.3$), b_i($i = 1.3$), c_i($i = 1.3$), d_j($j = 1.7$), k_m ($m = 1.8$), α_n($n = 1.6$), β_i ($i = 1.3$), δ_m($m = 1.8$), γ_m ($m = 1.8$) – model parameters.

Functions $E(t)$ – energy substrates, $Pr(t)$ – prostaglandins, $G(t)$ – $cGMP$, $A(t)$ – $cAMP$, $Lp(t)$ – 5-lipo-oxiarachidonic acids, $P_{Ca}(t)$ – calcium pump, $I(t)$ – iodine ion level in thyroid gland; $Ah(t)$ – hypothalamus activity; $C(t)$ – cortisol; $D_N(t)$ – dopamine, $IL1(t)$ – interleukin-1; $GH(t)$ – growth hormone, $S(t)$ – somatostatin, $ES(t)$ – estrogens can be defined based on the mathematical model (Atoyev 1999).

A software for NIEN analysis at some types of TS pathologies (hypo and hyperthyroidism, T3- and T4-thyreotoxicoses) has been developed; it has been used to assess the role of metabolic and thyroid components in a NIEN disturbances development process.

TH dynamics for norm, hypothyroidism, and hyperthyroidism, obtained in a model experiment, is shown at Fig. 6.14. The frequency of T3 and T4 oscillations and their amplitudes increase during hyperthyroidism according to the model. The opposite results were obtained in the case of hypothyroidism.

TH changes under T3- and T4-thyreotoxicosis have been also investigated. T3-thyreotoxicosis causes an increase of T3 and a frequency of its oscillations, but decrease of TTH. The oscillations of T4 practically do not vary on amplitude, but vary on a phase. T4-thyreotoxicosis causes an essential increase of the level of T4 and a frequency of its oscillations. A frequency of T3 oscillations also considerably increases, but their amplitude does not vary so intensely. A level of TTH decreases to a lesser degree than during T3-thyreotoxicosis.

Average values of metabolic factors in the interval of modeling are presented at Fig. 6.15. Basing on the model we can conclude that hyperthyroidism, T3- and T4-thyreotoxicosis are not linked to essential changes in metabolic factors dynamics. On the contrary, hypothyroidism is a unique disease (from the above list), which is linked with an amplification of Ca^{++} influx, intensification of cell metabolism, and essential changes of metabolic factors. First of all, those are CN and AA.

Our research results in the following ranking of thyroid pathologies by the role of metabolic impacts in NIEC disturbances: T4-thyreotoxicosis (minimal impact), T3-thyreotoxicosis and hyperthyroidism (medium), hypothyroidism (maximal). Thus, the following schemes of development of pathological processes may be considered.

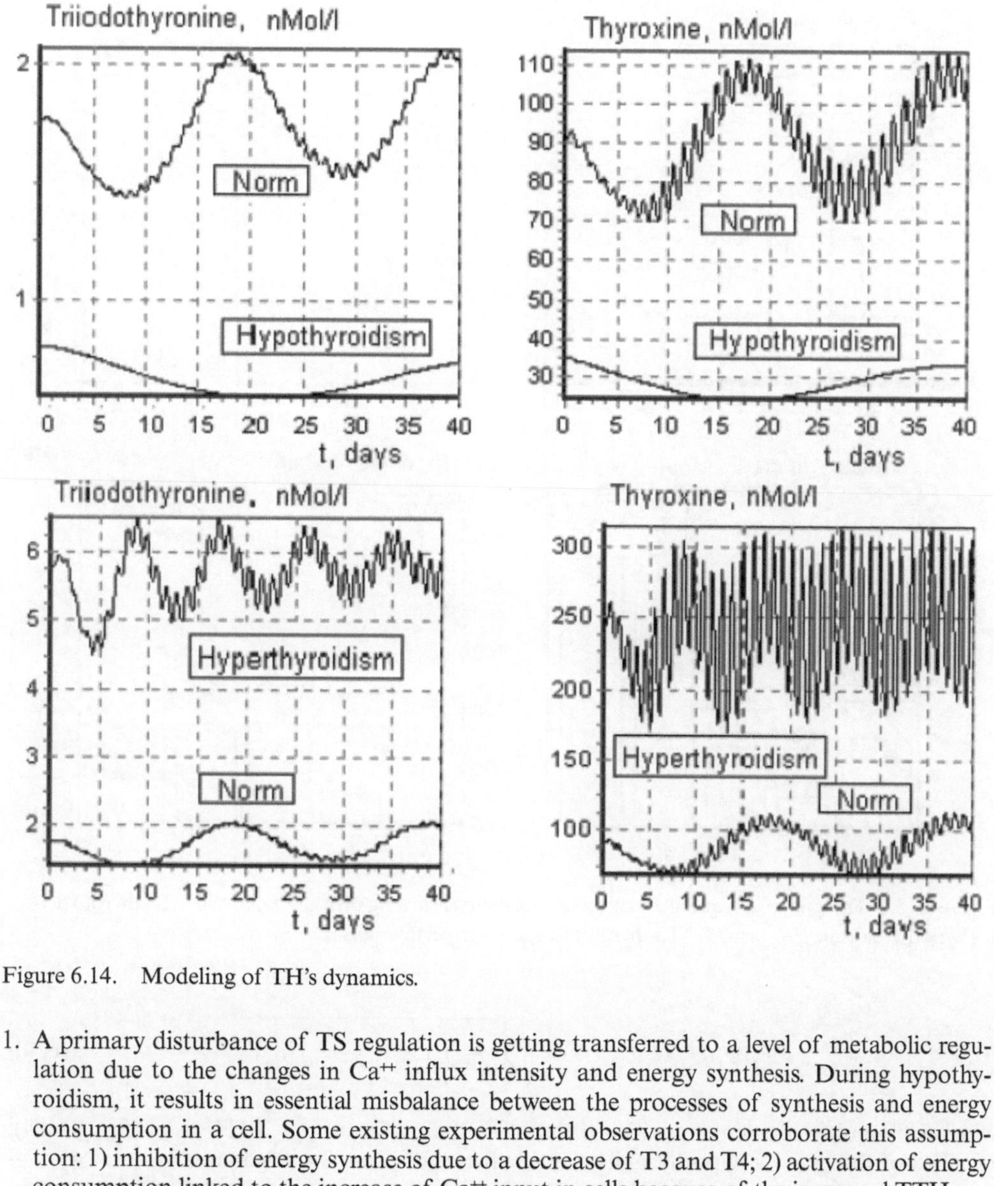

Figure 6.14. Modeling of TH's dynamics.

1. A primary disturbance of TS regulation is getting transferred to a level of metabolic regulation due to the changes in Ca^{++} influx intensity and energy synthesis. During hypothyroidism, it results in essential misbalance between the processes of synthesis and energy consumption in a cell. Some existing experimental observations corroborate this assumption: 1) inhibition of energy synthesis due to a decrease of T3 and T4; 2) activation of energy consumption linked to the increase of Ca^{++} input in cells because of the increased TTH.
2. The energy disbalance causes metabolic shifts, for instance cAMP and prostaglandins level changes. Probably, they have compensatory character to limit negative changes in TH ratio.

6.3.4 *Optimal control of the thyroid system restoration*

The task of optimal control of TS restoration has been defined as follows: mathematical model of TH synthesis, control parameters U_i ($i = 1$ – activators of oxidative phosphorylation, $i = 2$ – activators of Ca^{++} transport, $i = 3$ – Iodine ions, $i = 4$ – intensity of Ca^{++}-pump work, $i = 5$ – inhibition of hypothalamus activity by cortisol) and their minimum and maximum values, accordingly U_{imin}, U_{imax} have been given. Control effects and phase trajectories minimizing functional Fj on an interval $(0, T_k)$ are to be found:

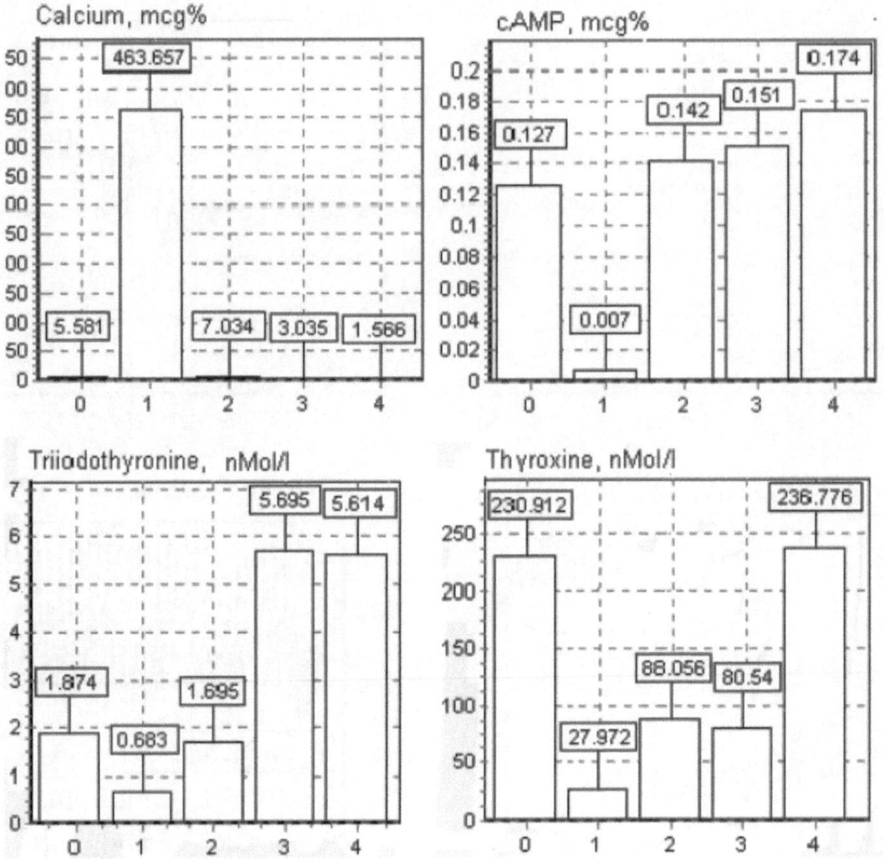

Figure 6.15. The comparative analysis of metabolic changes during thyroid diseases: 0 – T4-thyrotoxicosis, 1 – hypothyroidism, 2 – norm, 3 – T3-thyrotoxicosis, 4 – hyperthyroidism.

$$F_j(U_1,U_2,U_3,U_4,U_5)=\sqrt{\sum_{k=1}^{3}\gamma_{ik}\sum_{l}^{n}\left(x_{ik}(t_l)-x_{Hk}(t_l)\right)^2}, \tag{6.40}$$

where n – number of points in which the values x_{jk} were fixed (x_{j1} – TTH, x_{j2} – T3, x_{j3} – T4); j – index of disease (1 – hypothyroidism, 2 – hyperthyroidism, 3 – T3-thyreotoxicosis, 4 – T4-thyrotoxicosis); x_{Hk} – values of TH levels appropriate to norm; γ_{ik} – weight coefficients introduced to equalize a contribution of each added to the objective function, with the limitations $U_i \in [U_{imin}, U_{imax}]$. The method of casual search has been selected as an optimization method.

To define an efficiency of TS normalization, the quality of optimization has been introduced as an index (*QO*), showing in how many times the deviation of each TH from norm has been reduced in a result of control. The indices *QO* are defined as:

$$QO_{X\,jk}=\left|I_{X\,Hk}-I_{X\,jk}\right|/\left|I_{X\,Hk}-I_{X\,Optjk}\right|, \tag{6.41}$$

The following denotations of k-th hormone indices have been accepted:

$$I_{X\,Hk}=\int_0^T x_{Hk}\,dt, \tag{6.42}$$

$$I_{X\,jk} = \int_0^T x_{jk}\,dt, \tag{6.43}$$

$$I_{OptX_{jk}} = \int_0^T x_{jk}(U_i)dt. \tag{6.44}$$

They determine total amount of k-th hormone in the time interval [0, T] in norm, at disease j, and at presence of control effects U_i. The model experiments were carried out. Their objective was normalization of TH levels at the mentioned above diseases. The modeling results obtained at controls U_i ($i = 1.5$) have allowed us to rank the efficiency of control process under above diseases (Fig. 6.16).

Following are *QO* consequences in the descending order.

1. Hyperthyroidism. Small decrease of T3 and T4 levels and small increase of TTH level are possible. *QO* indices are, respectively, 1.55966, 1.72004, and 1.15142.
2. T3-thyreotoxicosis. The decrease of T3 and T4 and rise of TTH levels are insignificant. *QO* indices are, respectively, 1.07491, 1.04857, and 1.62824.

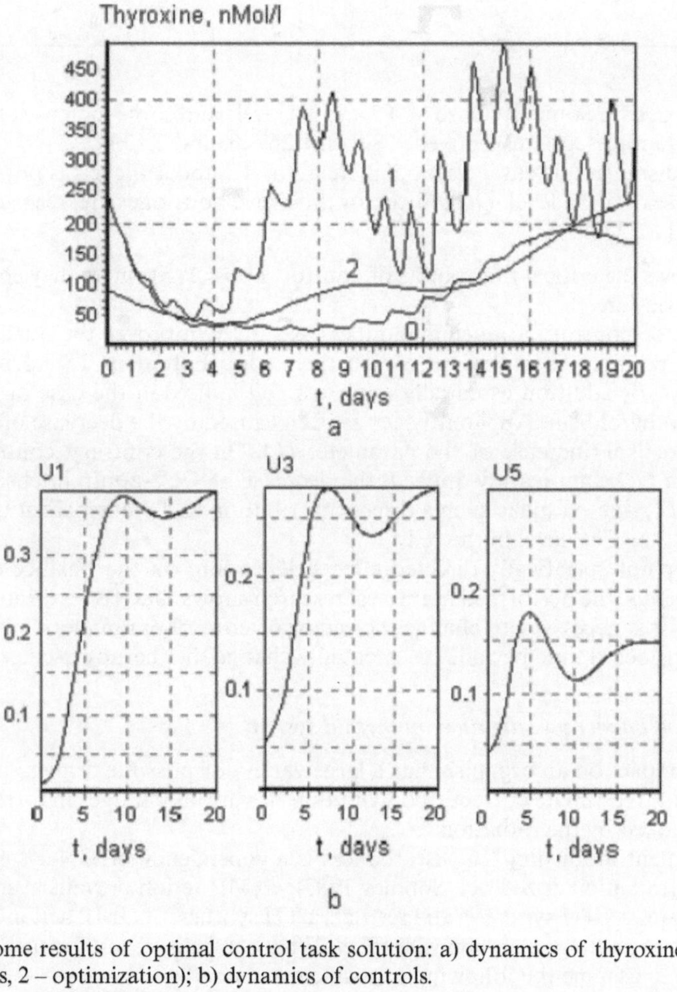

Figure 6.16. Some results of optimal control task solution: a) dynamics of thyroxine (0 – norm, 1 – T4-thyrotoxicosis, 2 – optimization); b) dynamics of controls.

Table 6.4. Changes of controls under restoration of thyroid gland (EN – extremum number, TI – time interval (hours), TC – type of changes: ↓ – decrease, ↑ – increase).

Disease		U_1	U_2	U_3	U_4	U_5
Hyper-thiro-idismo	EN	1	1	1	1	1
	TI	0–6, 6–20	0–8, 8–20	0–7, 7–20	0–5, 5–20	0–6, 6–20
	TC	↓↑	↓↑	↓↑	↓↑	↑↓
T3-thy-reoto-xicosis	EN	1	1	2	1	1
	TI	0–12, 12–20	0–6, 6–20	0–4, 4–12, 12–20	0–5, 5–20	0–8, 8–20
	TC	↓ ↑	↑↓	↑↓ ↑	↓↑	↑↓
T4-thy-reoto-xicosis	EN	1	1	1	1	1
	TI	0–10, 10–20	0–8, 8–20	0–14, 14–20	0–6, 6–20	0–8, 8–20
	TC	↑ ↓	↓↑	↑ ↓	↓↑	↑↓
Hypo-thiro-idism	EN	1	2	2	1	1
	TI	0–8, 8–20	0–5, 5–10, 10–20	0–7, 7–18, 18–20	0–10, 10–20	0–7, 7–20
	TC	↑↓	↓↑ ↓	↓↑ ↓	↑ ↓	↑↓

3. T4-thyreotoxicosis. Some decrease of T3 and T4 levels and some increase of TTH level are possible. *QO* indices are, respectively, 18.948, 1.26746, and 2.17923.
4. Hypothyroidism. In this case, essential increase of T3 and T4 levels is possible, as well as some decrease of TTH level. *QO* indices for the above hormones are, respectively, 24.9942, 2.52003, and 3.23769.

Table 6.4 shows the critical time points of controls U_i $(i = 1, 5)$ where they change the direction of their variation.

The increase of control parameters number essentially improves the quality of the optimization. It is possible to rank controls on their contribution in TS normalization. For instance, control U_5 addition essentially improves *QO* indices in the case of T3 normalization under hypothyroidism. Apparently, it has been caused by the decrease of hypothalamus inhibition by cortisol (increase of the parameter U_5). On the contrary, control U_4 addition doesn't improve *QO* significantly. Indeed, the decrease of Ca^{++}-pump intensity (increase of the parameter U_4) doesn't make such a direct contribution to TS normalization. In this case, there is a growth of Ca^{++} level in the cell.

It can render not specifically directed effects depending on the balance of various cell intermediates. Thus, the performed numerical research shows that TS restoration is a complicated problem. It is necessary to change the vector of control parameters' variations several times in the considered time period, i.e. essentially change the therapy used.

6.3.5 *Radiation impact investigation on thyroid system*

The radiation impact on an organism has a large variety of possible targets. This complexity makes it difficult to estimate correctly contribution of various changes in forming pathological processes evoked by the radiation.

The key moment in forming TS disturbances is a dependency of CN system (cAMP and cGMP) on an irradiation doze level (Sobolev 1987). cAMP action prevails at small irradiation dozes. It stimulates a TTH synthesis and restricts a TH synthesis. cGMP action prevails at high irradiation dozes, resulting in an opposite effect. What can be a reason for such a discrepancy?

It is possible to assume the following probable reasons:

1. Free radicals' accumulation at high irradiation dozes; it causes the decrease of a prostaglandin's synthesis rate, in spite of the increased level of Ca^{++} and weakening one of adenylatcyclase activation mechanisms.
2. Exhaustion of the adrenergic system also causes the decrease of an adenylatcyclase activity.
3. Allosteric regulation of phosphodiesterase activity by a high level of cGMP; calcium accumulation in the cell at high irradiation dozes also causes the rise of cGMP level. High level of cGMP increases the activity of phosphodiesterase causes acute drop of cAMP level, which cannot be compensated by a stimulation of the adenylatcyclase activity due to prostaglandins.

The model (6.39) has been utilized to investigate possible mechanisms of pathology forming caused by a radiation action on the organism. We took into account that an increase of the exposure doze (D) varies the rates of TH synthesis and destruction, as well the intensity of Ca^{++} ions fluxes through cellular membrane in both directions (see Fig. 6.13). Based on these data, let us introduce the functions to define a dependence of some model parameters on a radiation doze: $d_j = d_j(D)$, $(j = 1,7)$, $b_1 = b_1(D)$, $\beta_2 = \beta_2(D)$, $\gamma_2 = \gamma_2(D)$. These functions have been determined based on the parameter identification problem solution using the method of causal search.

The mathematical problem statement has been formulated as follows:

Let $X_i = (x_i(D_1),\dots\, x_i(D_n))$, $(i = 1,3)$ be a sample of experimental data, where X_1 – TTH, X_2 – T3, X_3 – T4, $x_i(D_n)$ is a level of i-th hormone in the middle of n-th intervals of exposed doses. Let the mathematical model be given as:

$$dx/dt = F\left(t, x, U_1(D) \dots U_1(D)\right), \tag{6.45}$$

where x(t) – vector of model variables, $U_m(D)$ – vectors of model coefficients depending on the dose.

It is required to find vectors $U_m(D)$ on the interval (t_0, t_n) to minimize the functional

$$I = \left\{ \sum_{i=1}^{3} G_i \sum_{j=1}^{n} \left[x_i(D_j) - \overline{x_i(D_j)} \right] \right\}^{1/2}, \tag{6.46}$$

where $x_i(D_j)$ and $x_i(D_j)$ are the model and sample levels of i-th hormone under the exposed dose from j-th interval, G_i-weight coefficients.

Table 6.5 shows a comparison of model results (MR) and experimental data (ED) obtained during an inspection of persons injured as a result of the Chernobyl accident. The dose is measured in Grey (Gy).

Fig. 6.17 shows the total amount of TH versus the irradiation doze. The results of a performed model research made it possible to rank sensitivity of different TS regulation links

Table 6.5. Irradiation doze impact on thyroid hormones level.

Hormones (nmole/l)	Data type	Norm	Exposed doze 1–2 Gy	2–4 Gy	10 Gy
T3	ED	1.77 ± 0.15	1.76 ± 0.25	1.57 ± 0.36	3.09 ± 0.23
	MR	1.77	1.75	1.56	2.48
T4	ED	91.34 ± 5.05	85.20 ± 7.36	128.24 ± 7.97	157.21 ± 10.01
	MR	92.48	91.78	127.61	148.24
TTH	ED	1.91 ± 0.20	2.09 ± 0.32	1.86 ± 0.32	1.52 ± 0.12
	MR	1.91	2.10	1.87	1.79

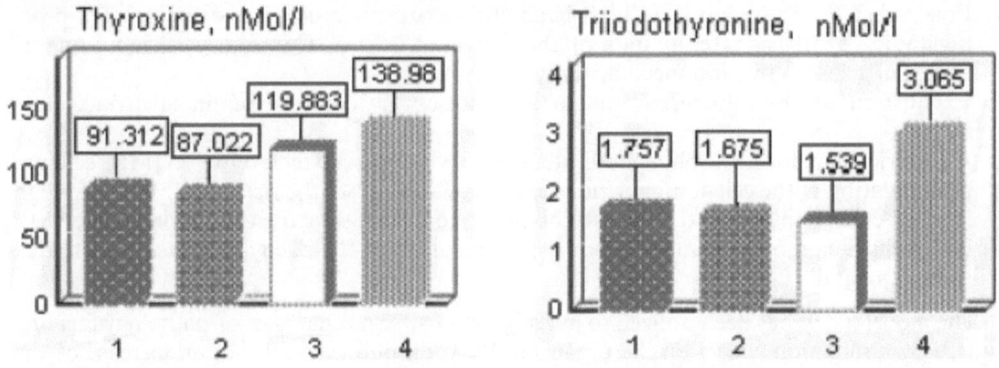

Figure 6.17. Model research of radiation impact on a total level of TH (columns show average values of hormones during time of modeling experiment: 1 – D = 0, 2 – D = 1.5 Gy, 3 – D = 3.0 Gy, 4 – D = 8.0 Gy).

to radiation action. At the dose of 0–2 Gy, the most influenced by radiation impact are the processes of TH synthesis, TTH, and TRF destruction, as well as Ca^{++}-pump. At the doze of 2–4 Gy, the radiation also affects a Ca^{++} influx. At the doses of 4–10 Gy, the radiation also affects a destruction of T3 and T4. It has been known that various cellular processes which were considered as targets of a radiation action are controlled by the CN system. Besides, it has been shown that a dysfunction of the CN system manifests in early periods after irradiation and can precede changes in other systems. All these facts raise the question whether postradiation changes in the CN system are the mechanism triggering disturbances in other systems. In this case the disturbances are caused rather by metabolic modifications in cells connected with the changes of the CN ratio, than by a direct effect of free radicals formed by irradiation and other products of radiolysis (at least, on the early stages).

Thus, the research of a CN role in formation of post-radiation disturbances is one of the central problems of modern radiobiology. The increasing number of publications makes the conclusion that this research can shed light on the mechanisms forming radial affection and permit a development of complex therapeutic schemes of the individualized therapy, as well as new methods for estimation of its efficiency and health risk. The mathematical modeling of such complex system has demonstrated possibility to get an approximation of the real sequence of the events developing in the organism as a whole under the action of radiation and others stress factors, on the basis on some integrated parameters of environmental impact on the organism as a whole and on its separate control systems. The research made it possible to rank various links of TS by their sensibility to a radiation impact and determine the following most probable sequence of their changes: radiation action $\rightarrow$ change in the NIEN system $\rightarrow$ change of ACTH and cortisol levels $\rightarrow$ change of TTH level $\rightarrow$ change of Ca^{++} input $\rightarrow$ change in the CN system $\rightarrow$ change in TH synthesis $\rightarrow$ disturbances in metabolism of Ca^{++} ions.

6.4 CONCLUSIONS

A renewed focus on inequality and sustainability characterizes the great part of scenarios elaborated by the UN experts during last decade. Their common idea is the following: growing inequality reduces opportunities of sustainable development whereas greater equality makes societies stronger. The challenges of global changes turn environmental and social protection systems to necessary components of survival in XXI century. Such conclusions become more apparent for developing world, including Ukraine. For Eastern Europe, the reforms of social sphere and system of energy, water and food consumption may become driving forces that accelerate its development and reduce lag behind leaders demonstrated their abilities to implement the reforms.

However, the reforms are not viable without appropriate facilitating tools. The information systems based on mathematical models are important components of such tools. They improve essentially outcomes of reforms and foster timely decision makings on environmental, financial, economic, and social issues. Those information systems also provide scientific and technological basis to develop scenarios for society safe in various dimensions, speed-up of sustainable economic growth, and increase in life quality.

The main objective of this scientific research is to promote new approaches for sustainable management of the environment through developing new technologies of vulnerability and risk assessment and management for the complex environmental systems for which standard modeling methods and tools are inadequate.

The practice of the approach to risk analysis describes herein allows not only for estimation of the risk of emergency, but also for the calculation of the quantitative characteristics of reserve possibilities of complex system and its components. It also allows for the description of the current state of the system by ranking a set of risks of emergency occurrence in its separate links, thereby identifying the weakest link and area of needed improvement. The main advantage of this approach is the determination of vulnerability as the function of dynamic variables of the investigated ecological, social, technogenic, economic and biological systems.

Proposed approach may be transformed into the baseline tool for increasing of horizon of forecasting, also as for elaborating the global strategy of disaster and other extreme and rare events prevention, which the world currently lacks.

REFERENCES

Anderson J.J., 1985: A theory for attitude and behavior applied to an election survey. *Behavioral Sciences 30*, pp. 219–229.

Atoyev, K.L. and Rykhtovsky V.O., 2000: Computer technology for health risk estimation and management. Proceedings of ESREL 2000 and SRA-Europe annual conference "Foresight and Precaution". Ed. J. Tait. *Belcema Publishers*, Rotterdam, Netherlands, pp. 109–115.

Atoyev, K.L., 2002: The integral assessment of bioterrorism threat. Supercourse: Epidemiology, the Internet and Global Health. Pittsburgh: Disease Monitoring and Telecommunication WHO Collaborating Center, www.pitt.edu/~super1/lecture/lec6221/index.htm.

Atoyev K.L., 1999: Investigation of the influence of economic, financial, political and social factors on risk of social shocks. Proceedings of the 9th Annual Conference Risk Analysis: facing New Millennium. Ed. L.H.J. Goossens, *Delft University Press*, Netherlands, pp. 616–620.

Atoyev K.L., 1999: Research of role of metabolic and hormone mechanisms of neuro-immune-endocrine regulation in organism's reactions on stress. Proceedings of the XXIII International Workshop on Modelling of Developing Systems. *Glushkov Institute of Cybernetics*, Kyiv, pp. 94–100.

Atoyev K.L., 2001: Risk Assessment in Ukraine: New Approaches and Strategy of Development. Assessment and management of environmental risks: methods aid applications in Eastern European and developing countries, Ed. I. Linkov and J.M. Palma Oliveira, *Kluwer*, Amsterdam, pp. 195–202.

Atoyev K.L., 2002: New Approaches to Epidemiological Risk Assessment Management. Supercourse: Epidemiology, the Internet and Global Health. Pittsburgh: Disease Monitoring and Telecommunication WHO Collaborating Center, http://www.pitt.edu/~super1/lecture/lec4871/index.htm.

Atoyev K.L., 2005: Mathematical modeling of metabolic and hormonal regulation: risk assessment of environmental and radiation influence on various links of endocrine system. *HAIT Journal of Science and Engineering B 2(1–2)*, pp. 31–53.

Atoyev K.L., 2007: Multicriteria Decision-Making Tool for Optimal Water Resourse Management. Water Reuse-Risk Assessment Decision-Making and environment, Ed. M.R. Zaidi, *Springer*, Dordrecht, pp. 131–144.

Atoyev K.L., 2015: A comprehensive evaluation of investment attractiveness of agricultural regions in the face of increasing uncertainity and risk. *Mathematical modelling in economy 1*, pp. 96–106, (in Russian).

Atoyev K.L., Bilous V.M., Deineka V.S., 2013: Mathematical modelling of food, energy and water resources management under increased uncertainties and risks. Integrated modeling of food, energy and water security management for sustainable soial, economic and environmental developments. Ed. A.G. Zagorodny and Yu. M. Yermoliev, Kiev, pp. 195–228.

Atoyev K.L., Golodnikov A.N., Gorbacuk V.M., Ermoliev Yu.M., Ermolieva T.Y., Kiriljuk V.S. and Knopov P.S., 2014: The mathematical problems of complex management and effective utilization of food, energy and water resources under increased uncertainties. Integrated Management, Security, and Robustness. Ed. A.G. Zagorodny, Yu.M. Yermoliev and V.L. Bogdanov, Kuiv, pp. 198–227.

Atoyev K.L., Tomin A. and Aksionova T., 2007: Global changes, new risks, and novel methods and tools of their assessment. Modeling and management of environmental security in Ukraine, "Managing Critical Infrastructure Risks", Ed. I. Linkov et al., *Springer*, Netherlands, pp. 339–351.

Capra F., 1996: The Web of Life: A New Scientific Understanding of Living Systems. *Anchor*, N.Y.

Capra F., 2002: The Hidden Connections: A Science for Sustainable Living. *Random House*: New York.

Cellier F.E., 2008: World3 in Modelica: Creating System Dynamics Models in the Modelica Framework. Proceedings of the 6th International Modelica Conference. *Bielefeld*, Germany, pp. 393–400.

Cobb L. and Zacks S., 1985: Applications of Catastrophe Theory for Statistical Modeling in the Biosciences. *J. Am. Stat. Assoc. 80*, pp. 793–802.

Ermoliev Y. and von Vinterfeldt D., 2012: Systemic risk and security management. Managing Safety of Heterogeneous Systems, Lecture Notes in Economics and Mathematical Systems. *Springer-Verlag*, Berlin, Heidelberg, pp. 19–49.

Forrester J., 1978: World Dynamics. Moscow, Nauka, 167, (in Russian).

Gelovani V.A., Egorov V.A., Mitrofanov V.B. and Piontkovskiy A.A., 1974: Solution of One Control Problem for Forrester Global Dynamical Model. Moscow, preprint-56, (in Russian).

Gelovani V.A., Kurakin P.V., Malinetskii G.G. and Makhov, S.A., 2005: Steady-state solutions in a modified Forrester model. *Dokl. Math. 71(2)*, pp. 186–188, (in Russian).

Guastello S.J., 1998: Catastrophe Modeling of the Accident Process: Organizational Subunit Size. *Psychol. Bull. 103*, pp. 246–255.

Kahnemann D. and Tversky A., 1972: Subjective probability: a judgement of representativesness. *Cognitive Psychology 3*, pp. 430–454.

Klinke A. and Renn O., 2002: A new approach to risk evaluation and management: risk-based precaution-based and discourse based strategies. *Risk Analysis 22(6)*, pp. 1071–1094.

Knopov P.S. and Derieva O.N., 2013: Estimation and Control Problems for Stochastic Partial Differential Equations. *Springer*.

Knopov P.S. and Korkhin A.S., 2012: Regression Analysis Under a priory Parameter Restrictions. *Springer*.

Knopov P.S. and Pardalos P.M., 2009: Simulation and optimization methods in risk and reability theory. *Nova Sciencie Publishers Inc*, New York, pp. 285.

Korotayev A. and Tsirel S., 2010: A spectral Analysis of World GDP Dynamcs: Kondratieff Waves, Kuznets Swings, Juglar and Kitchin Cycles in Glibal Economic Development, and the 2008–2009 Economic Crisis. *Structure and Dynamics 4(1)*, pp. 3–51.

Lange R., Oliva T.A. and McDade S.R., 2000: An Algorithm for Estimation Multivariate Catastrophe Models: GEMCAT II. *Studies in Nonlinear Dynamics and Econometrics 4*, pp. 137–168.

Lapushenko O., Atoyev K.L., Berejnov S., 2004: The Estimation of HIV/AIDA escalation in Ukraine: Information Technology for Management of Epidemics and risk ranking. *Vrachebnoe delo 5/6*, pp. 3–17, (in Russian).

Laszlo F., 1993: The Creative Cosmos. *Floris Book*, Edinburg.

Lorenz E., 1963: Deterministic nonperiodic flow. *Journal of Atoms. Sci. 20*, pp. 130–141.

Magnitskii N.A. and Sidorov S.V., 2006: New Methods for Chaotic Dynamics. *World Scientific*, 380.

Makhov S.A., 2007: Mathematical simulation of world dynamics and sustainable development by the example of Forrester's model. Self-organizing systems. Moscow, *Nauka*, pp. 79–101, (in Russian).

Marcuk G.I., 1986: Mathematical Models in Environmental Problems. *Elsevier Science Pub. Co Elsevier*, Amsterdam, 216.

Moiseev N.N., Alexandrov V.V. and Tarko A.M., 1985: Human and Biosphere, Moscow. *Nauka*, 272, (in Russian).

Moskalenko F., Atoyev K.L., Berejnov. S., 2002: Epidemiological risk management in Ukraine: new approaches and strategy of development. *Vrachebnoe delo 3/4*, 1–13, (in Russian).

Myasnikov A.A., 2011: Synergetic effects in modern economy: introduction in a problematics. *LENAND* Moscow, 160, (in Russian).

Pepeliaev V.A., Knopov P.S., Atoyev R.L., Bigdan V.B., Tchortnyy Y.M., 210: Information-analytical system for complex environmental-technogenic and social-economic risks analyses in the field of housing and communal service in Ukraine. *Science and innovation 6(3)*, 39–46, (in Ukrainian).

Poston T. and Stewart I., 1978: Catastrophe Theory and its Applications. *Dover*, New York.

Prigogine I. and Stengers I., 1984: *Order out of Chaos*, Bantam, N.Y.

Puu T., 2000: Attractors, Bifurcatorions, and Chaos—Nonlinear Phoenomena in Economics, *Springer*.

Renn O., 2003: Typology of risks. *Journal of risk research 6(4–6)*, pp. 519–524.

Sergienko I.V., Yanenko V.M. and Atoev K.L., 1995: Optimal control of the immube response synchronizing the various regulatory compartments of the immune system. I. Mathematical analysis of the risk of pathological loaders in organism. *Cybernetics and Systems Analysis 31(2)*, 225–239.

Sergienko I.V., Yanenko V.M. and Atoev K.L., 1997: A conceptual framework for managing the risk of ecological, technogenic, and sociogenic disasters. *Cybernetics and Systems Analysis 33(2)*, pp. 203–219.

Shvitra D., 1989: Dynamics of Physiological Systems. Vilnius, *Mokslas*, 169, (in Russian).

Sobolev A., 1987: Radiation Biochemistry of Cyclic Nucleotides. Moscow. *Nauka*, 179, (in Russian).

Thabrew I., Ayling R.M. and Wicks C., 2004: Biochemistry for Clinical Medicine. *Cambridge Univ. Press*, Cambridge.

Thom R., 1975: Structural Stability and Morphogenesis. *Benjamin-Addison Wesley*, New York.

van der Maas H.L.J., Kolstein R., Van der Pligt J., 2003: Sudden Transitions in Attitudes. *Soc Meth. & Res. 32(2)*, pp. 125–152.

Wagenmakers E.-J., Peter C.M., Molenaar P.C.M., Grasman R.P.P., Hartelman P.A.I. and Van der Maas H.J.L., 2005: Transformation invariant stochastic catastrophe theory. *Physica D 211*, pp. 263–276.

Wilkinson A., Elani S. and Eidinow E., 2003: Background and dynamics of the scenarios. *Journal of risk research 6(4–6)*, pp. 365–401.

Wilkinson A., Elani S. and Eidinow E., 2003: Riskworld scenarious. *Journal of risk research 6(4–6)*, pp. 297–334.

Zhang W.-B., 2005: Differential Equations, Bifurcations, and Chaos in Economics, *World Scientific publisher.*

Zulpukarov M.-G.M., Malinetskii G.G. and Podlazov A.V., 2006: The channels and joker's method application to the Rosenzweig–Macarthur system analysis. *MKO 2*, 28–38, (in Russian).

CHAPTER 7

Study of the impact of the technical state of the transformers with the LTC on the parameters of the EES modes optimal control

O.E. Rubanenko & O.I. Kazmiruk
Vinnytsia National Technical University, Vinnytsia, Ukraine

T. Zyska & K. Gromaszek
Lublin University of Technology, Lublin, Poland

M. Junisbekov
M.Kh. Dulaty Taraz State University, Taraz, Kazakhstan

INTRODUCTION

Characteristic feature of a present day situation is that energy companies attempt to improve energy efficiency in conditions of aging of high voltage power equipment that is involved in the processes of the optimal control at the stages of generation, transport and consumption of electric energy.

Large-scale introduction of methods and means of intellectual support of optimal decision-making proves their efficiency. One of the directions of efficiency increase in electric energy transportation is improvement of methods and means of active power losses reduction on conditions of maintaining reliable operation of high voltage equipment, including out-dated equipment.

At any moment of time, energy system is in the state that is determined by its parameters. The set of energy system states and processes the transition from one state into another. Its mode is characterized by the parameters, for instance electrical: voltages and substations loading, current in the transmission lines, transformation ratio of the transformers, etc.

Normal operation of energy systems is only possible in conditions of reliable functioning of a real-time operation control system of EES modes. Modern technologies of optimal control provide the improvement of methods and facilities of their information support. Concept of Smart Grids provides large-scale implementation of energy saving technologies in conditions of outdated equipment operation.

Jakushokas (2013) suggested the technique, aimed at reducing of voltage drop in separate parts of distribution networks, however it does not take into account a technical state of regulation devices. Tirupathi Reddy (Tirupathi et al. 2012) considers the example of electric networks in India, statistics of power value increase that is transported in electric network. It is proved that such increase leads to the overloading of certain transmission lines, to rapid aging of high voltage equipment, undesirable rapid reduction of residual operation life of the equipment, to its failures. As a result, duration and cost of damaged equipment repair increases, power losses grow in the branches of electric systems, which in the period of repair operate in non-optimal modes. It is stated that for the increase of network reliability and efficiency of energy transmission, it is necessary to use power transformers, equipped with load changing devices and automatic or automated control systems of such LTC devices, that would control power flows in the branches of the circuit and perform the control oven LTC in such a way that the parameters of electric network modes were within the limits of technical parameters of the equipment (transmission lines, switching devices, transformers, etc). It is provided by usage of FACTS technologies.

M. Kolcun, D. Hlubeň, L. Beňa, N. Djagarov and Z. Grozdev (Kolcun et al. 2010) consider the possibility of using of phase-shifting transformer for change of power flows in an electric energy system in order to decrease electric power losses during its transmission in the lines of Slovak Republic. M. Bocovich, K. Iyer, R. M. Terhaar, N. Mohan (Bocovich et al. 2013) show high cost of FACTS technologies usage in power industry for reduction of electric power losses.

C. Constantin, M. Eremia, L. Toma (Constantin et al. 2013) consider three variants of power flows control in electric network, using three FACTS devices: with Thyristor Controlled Series Capacitor (TCSC), Static Synchronous Series Compensator (SSSC) and phase-shifting transformer (PST), but the state of equipment that performs modes control, is not paid attention to.

In (Alekseev 2010) B.A. Alekseev, on the base of the analysis of literature sources carried out, makes a conclusion that prolongation of power transformers lifetime for the period of 20–30 years is more profitable that their replacement by new ones; the amount of power transformers in the USA that operate more than 25 years (certificate resource – 25 years) is about 65%. That is why computer systems that perform non-stop monitoring over the state of high voltage equipment and its nodes function (Rassalsky 2010).

The example of such systems may be «КИН-750» or «КИВ-500» that enable us to control the state of insulation of power transformers bushings during operation and other equipment. As B.S. Stogniy, M.F. Sopel state (Stogniy and Sopel 2013), control of high voltage equipment provides the realization of tasks dealing with the monitoring of the values of its diagnostic parameters and determination of current technical state of this equipment. As B.S. Stogniy, O.V. Kirilenko, O.F Butkevych, M.F. Sopel state (Stogniy et al. 2009) basic technical characterizes and indicates modern monitoring facilities that enable us to improve the reliability of a real-time operation control system of United Energy System (UES) of Ukraine due to new possibilities of real-time operation control problems solution that appear during the introduction of such facilities.

The results of power transformer technical state determination may be used to define its current loading ability (Buslavets et al. 2015). In (Kylymchuk et al. 2015) it is proved that the application of new technologies, methods and means of modes control allows to reduce the losses of active power in electric networks. Taking into account the current state of power, transformers with on-load top changing devices decrease the risk of their damage during modes regulation (Lezhnyuk et al. 2012).

The aim of the research given is the development of the method of on-line diagnostics of the transformers with LTC and taking into consideration its results during optimal control of EES modes in order to minimize total losses of active power and excessive expenses of financial resources for the repair of the transformers in case of their failure (Lezhnyuk et al. 2012).

7.1 MATERIALS AND RESULTS OF THE RESEARCH

Automation of the process of power flow control may be provided by means of centralized remotely controlled alternative usage of switching devices (LTC) of the transformers. Under such conditions, the possibility of the analysis of control actions of separate LTC on mode parameters of EES appears by means of the feedback. This approach improves the operation quality of adaptive control automatic systems of the LTC position control. For this purpose, at considerable changes of load schedule it is necessary to perform ranking of the transformers with LTC by the quality of their impact on maintaining optimal parameters of the modes.

Realization of measures, aimed at reduction of power losses is limited by the possibilities of the equipment involved in the provision of optimal mode; namely, by its technical state. It is known that the damage of high voltage equipment during mode control (for instance, power transformers) leads to losses that considerably exceed the cost of electric energy, saved as a result of losses decrease.

Failure rate of the out-dated high voltage equipment (power transformers, shunting reactors, instrument current and voltage transformers, switches, etc.) increases when such equipment is in operation for more than 25 years (Tirupathi et al. 2012).

Taking into consideration the fact that the control of EES modes is accompanied by the operation of switching devices, regulation devices, transformers, switching surges, ferro-resonances, currents increase in power and instrument transformers, transmission lines, etc, then the control of modes must be realized, taking into consideration their technical state (Kolcun et al. 2010, Bocovich et al. 2013) and possible expenses for their replacement or repair.

Thus, it is necessary to know current state of high voltage electric equipment of EES, that is in operation during modes control.

7.2 DETERMINATION OF CURRENT TECHNICAL STATE OF POWER TRANSFORMERS

We will consider the methodology of determination of high voltage equipment current state on the example of power high voltage transformers, that have on-load-tap changing device. We suggest to evaluate technical state of power transformer by means of the analysis of the coefficient of its residual resource in the range from one (the best technical state) to zero (the worst technical state, when the transformer must be removed out of service for inspection, repair, replacement etc.).

Then, we will consider the example of residual resource coefficient determination of the transformer АТДЦТН 125000-330/110. First, we will study the statistics of failure rate of such transformers.

It is seen in Table 7.1 that, as a rule, the transformers are removed out of service as a result of humidity and contamination of oil, insulation as well as defects of bushings. The problem of development of mathematical model is complicated in conditions of incompleteness of initial data, when part of the parameters are known at the time of calculations, for instance, due to the necessity to carry out additional studies. To establish the interrelations of diagnostic parameters fuzzy modeling is very constructive technology. Such modeling allows us to obtain more reliable results as compared with the results of existing diagnostic systems.

Table 7.1. The reasons of removing power transformers out of service.

Transformer element	Designation	Parameter name	Parts	%
Windings	Z_k	Winding deformation	8	1.6
	t^o	Deterioration of contact joints state	10	2
	$P_{i,p}$	Idle power that characterizes of the magnetic quality	15	3
Insulation	R_{in}	Contamination of isolation	65	13.4
	W	Humidification of the isolation	48	10
Bushings	$k_{resid.res.bush}$ or k_{bush}	Defects of bushings	74	15.2
Oil	$CADG_c$	Content of dissolved gases	71	14.6
	PCA	High moisture content and deviations of other parameters of the oil	43	9
	$CADG_d$	Discharges in oil	64	13.2
LTC	$k_{def.LTC}$	LTC defects	45	9.3
Cooling system	$I_{motor.}$ or I_{mt}	The current of oil pump drive motor	14	2.9
	t^o_{cool}	Coolers temperature	16	3.3
Tank	k_{tank}	Tank leakage	12	2.5
Total			485	100

Symbols used in Table 7.1: Z_k – the resistance of the transformer windings (during measurements in short – circuit mode); t° – the temperature of contact points (for instance, bushing of the bus duct or with winding lead); $P_{i.p.}$ – idle mode power that characterizes the quality of magnetic circuit; R_{in} – the resistance of the insulation for revealing the contamination and aging of solid and liquid insulation (it is also necessary to determine the capacity and dielectric loss tangent, it is also desirable to determine the degree of polymerization); W – humidity of the isolation; $k_{resid.res.bush.}$ – the residual resource coefficient of the bushings; $CADG_C$ – residual resource coefficient of the transformer by the results of chromatographic analysis of dissolved gas in the transformer oil of the tank and LTC (ethylene, ethane, methane) of the transformer that characterizes oil contamination by the gases dissolved in it, for instance acetylene and hydrogen (for revealing of discharges); PCA – residual resource coefficient of the transformer by the results of physical-chemical analyses of transformer oil from transformer tank, contactor and LTC tap changer; $CADG_d$ – residual resource coefficient of the transformer by the results of chromatographic analysis of the dissolved hydrogen and acetylene in the transformer oil of the tank and LTC of the transformer tap changer in order to reveal the discharge; k_{LTC} – coefficient of the transformer LTC residual resource; I_m – the current of electric motors oil pumps and fans of cooling system; $t°_{cool}$ – coolers temperature; $k_{resid.res}$ – the residual resource coefficient of the transformer tank, determined by the availability (takes the value «0») or absent of oil leakage (takes the value 1).

Table 7.1 shows that the transformers are often displayed in repairs due to moisture and oil contamination, insulation and high-voltage inputs defects.

The task of creating a mathematical model complicated with incomplete initial data as part of the parameters known at the time of payment, such as the reasons for the need for additional studies. To establish reciprocal links diagnostic parameters very constructive simulation technology is unclear. This simulation allows us to obtain more reliable results compared to the results of existing diagnostic systems.

In Table 7.1, the controlled diagnostic parameter means deviation from which the norm helped to remove the transformer out of service or which was taken into account in the process of its removal out of service. In Table 7.1, the following diagnostic parameters are given: parameters, that characterize the state of the windings, insulation, bushings, oil, LTC, cooling systems, tank.

Having analyzed the data of the Table 7.1, the scheme showing whether the impact of diagnostic parameters is dependent or independent on the coefficient of total residual resource of the transformer (Fig. 7.1) was created.

Fig. 7.1 does not show mutual impact of one controlled diagnostic parameter on the other one. It is shown either in dependent or independent manner how these parameters influence the coefficient of total residual resource of Power Transformer (PT).

In Fig. 7.1, over the parameter the percentage amount of revealed faulty transformers by the given parameter is shown, that is given in percent of the total amount of faulty transformers. Consecutively denoted blocks with the deviations from which the norm helped to remove the transformers out of service. Blocks with the parameters that help to remove PT out of service on condition of simultaneous deviation of these parameters from the norms,

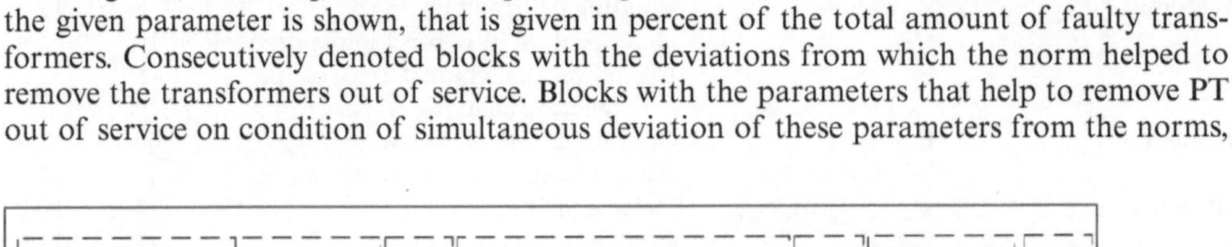

Figure 7.1. Structural diagram of the model of total residual resource coefficient of the transformer.

that is stipulated by the requirements regarding the reliability of transformers operation are shown in parallel (for instance, currents of electric motors of oil pumps and fans).

In order to obtain generalized index of residual resource of the transformer that takes into consideration the values of diagnostic parameters and their impact, it is suggested to pass to corresponding coefficients of residual resource by each diagnostic parameter from known values of diagnostic parameters (set in the given values). These coefficients are defined in relative units by (7.1) and that is why they characterize total output of the transformers from the moment of their technical state control to transition to boundary state that is residual technical resource (7.12). Residual resource coefficient k_i by i_1th diagnostic parameter:

$$k_{i_1} = \left| \frac{x_{i_1,lim} - x_{i_1,cur}}{x_{i_1,lim} - x_{i_1,in}} \right|, \tag{7.1}$$

where: $x_{i_1,lim}$ – is admissible limit normative value of i_1th diagnostic parameter; $x_{i_1,car}$ – is value of i_1th diagnostic parameter at the moment of control; $x_{i_1,in}$ – is initial value of i_1th diagnostic parameter (at the moment of putting into operation of new equipment or after repair), i_1 – is number of diagnostic parameter.

We perform the reduction of the circuit by the following expressions. For serial part of the circuit (Fig. 7.1), the coefficient of total residual resource is found by the following expression:

$$k_{tot.resid.res.} = \prod_{\tau=1}^{v} k_{\tau}^{p_{\tau}} \tag{7.2}$$

where: k_τ – is the coefficient of residual resource of PT by τth diagnostic parameter; τ – is τth diagnostic parameter; ν – is the amount of blocks in the serial part of the circuit in Fig. 7.1, p_τ – is the probability of controlled parameter deviation from admissible limiting normalized value of this parameter:

$$p_\tau = \frac{y_\tau}{m_2}, \tag{7.3}$$

where: y_τ – is a number of controlled parameter deviations from admissible limiting normalized value of this parameter, which were revealed by means of τth diagnostic parameter control (τ – for serial part of the circuit) from the total number of the revealed deviations of controlled parameters from admissible limiting normalized value; m_2 – is total quantity of the revealed deviations of controlled diagnostic parameter from their admissible limiting normalized values.

For parallel part of the circuit, the coefficient of total residual resource is found by the following expression:

$$k_{tot.resid.res.} = 1 - \sum_{j=1}^{m_1} \left[\left(1 - k_{res,j}\right) p_j \right], \tag{7.4}$$

where: $k_{res,j}$ – is the coefficient of residual resource of PT by j^{th} diagnostic parameter; j – is number of jth diagnostic parameter; m_1 – is a quantity of blocks (parameters) in parallel part of the circuit that is reduced.

The coefficient of total residual resource of PT is determined by the following expression:

$$k_{res} = k_{wind.} \cdot k_{in.} \cdot k_{bush} \cdot k_{oil} \cdot k_{LTC} \cdot k_{cool} \cdot k_{tank} \tag{7.5}$$

where: k_{wind}, k_{in}, k_{bush}, k_{oil}, k_{LIC}, k_{cool}, k_{tank} – are known at the moment of calculation of values of the coefficient of residual resource by the respective elements of the transformer.

7.2.1 *Neuro-fuzzy model of residual resource coefficient*

To create a mathematical model of residual resource, coefficient of the transformer parameters were used, by each of these parameters the conclusion regarding the state of the transformer can be made. But none of these parameters completely characterizes technical state of the transformer. It only shows certain changes of technical state of power transformer. Mathematical model of residual resource coefficient of the transformer was created by means of MatLab (Lezhnyuk et al. 2012). Using this model, it is possible to edit the already created (by the formula (7.5)) probabilistic sample of teaching data. These data help to obtain analytical dependence of residual resource coefficient of the transformer on diagnostic parameters in the form of the polynomial. For seven input parameters of the model that randomly changed from 0 to 1, the coefficient of total residual resource of the transformer (7.5) was determined, where input parameters of the model were reduced to relative units of their deviation from the norm.

By means of Anfis Editor using hybrid teaching algorithm and applying Sougeno algorithm of fuzzy conclusion (Kolcun et al. 2010) neuro-fuzzy model of residual resource coefficient of the transformer (using subclusterization method) was obtained.

The structure of the obtained neural network is shown in Fig. 7.2. For each input variable of neuro-model four linguistic terms with Gaussian membership functions (Kolcun et al. 2010) were used:

$$k_{res.i_1} = f\,(x_{i_1};\sigma_{i_1};c_{i_1}) = e^{\frac{-(x_{i_1}-c_{i_1})^2}{2\cdot\sigma_{i_1}^{\;2}}} \tag{7.6}$$

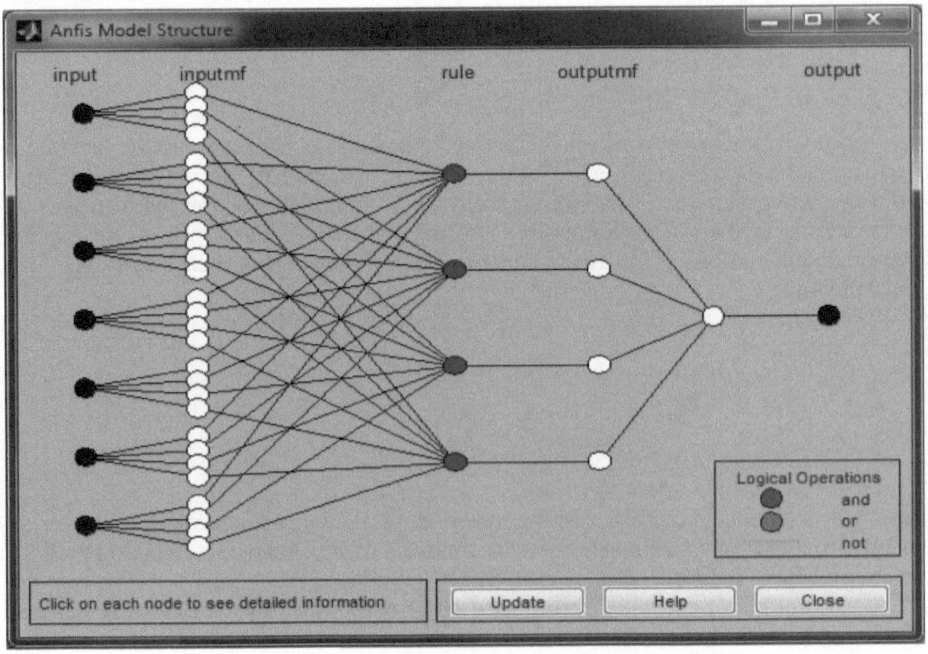

Figure 7.2. Structure of Anfis-network of the transformer.

where: δ_{i_1} and c_{i_1} – are numerical parameter; $\delta_{i_1}^2$ – in probability theory it is called dispersion of the distribution (7.14) and the second parameter c_{i_1} – is mathematic expectation; i_1 – is input parameter of neural-fuzzy model, that corresponds to diagnostic parameter (i_1 = 1, 2, 3, 4, 5, 6, 7), x_i – is value of i_1th input parameter of the model: $x_1 - k_{wind}$, $x_2 - k_{in}$, $x_3 - k_{bush}$, $x_4 - k_{oil}$, $x_5 - k_{LTC}$, $x_6 - k_{cool}$, $x_7 - k_{tank}$.

These are such terms as: "normal" values of diagnostic parameter, "minor deviations" of diagnostic parameter value, "prefault" values of diagnostic parameter, "emergency" value of diagnostic parameter.

For determining the value of total residual coefficient fuzzy non-linear autoregressive model of the total residual resource coefficient of the transformer is used. This model establishes fuzzy non-linear transformation between the values of residual resource coefficients by diagnostic parameters and total residual resource coefficient of the transformer:

$$k_{tot.resid.res} = F(k_{wind.} \cdot k_{in.} \cdot k_{bush} \cdot k_{oil} \cdot k_{LTC} \cdot k_{cool} \cdot k_{tank}) \tag{7.7}$$

where: F – is fuzzy functional transformation; $k_{wind.}$ – coefficient of residual resource of the windings; $k_{in.}$, $k_{bush.}$, k_{oil}, $k_{LTC.}$, $k_{col.}$, $k_{tank.}$ – coefficients of residual resource of the insulation, of bushings, of the oil, LTS, of system of the coling, of tank of the transformer.

$$\begin{cases}
\text{IF } k_{wind.} \in \text{“normal” AND } k_{in.} \in \text{“normal” AND } k_{Bush.} \in \text{“normal”} \\
\text{AND } k_{oil} \in \text{“normal” AND } k_{LTC} \in \text{“normal” AND } k_{cool.} \in \text{“normal”} \\
\text{AND } k_{tank} \in \text{“normal” THEN} \\
k_{tot.resid.res} = a_{11} \cdot k_{wind.} \cdot a_{12} \cdot k_{in.} + a_{13} \cdot k_{bush.} + a_{14} \cdot k_{oil} + a_{15} \cdot k_{LTC} + a_{16} \cdot k_{cool.} + a_{17} \cdot k_{tank} + c_1 \\
\text{IF } k_{wind} \in \text{“minor deviations” AND } k_{in} \in \text{“minor deviation”} \\
\text{AND } k_{BB} \in \text{“minor deviation” AND } k_{oil} \in \text{“minor deviation”} \\
\text{AND } k_{LTC} \in \text{“minor deviation” AND } k_{cool} \in \text{“minor deviation”} \\
\text{AND } k_{tank} \in \text{“minor deviation” THEN} \\
k_{tot.resid.res} = a_{21} \cdot k_{wind.} + a_{22} \cdot k_{in.} + a_{23} \cdot k_{bush.} + a_{24} \cdot k_{oil} + a_{25} \cdot k_{LTC} + a_{26} \cdot k_{cool.} + a_{27} \cdot k_{tank} + c_2 \\
\text{IF } k_{wind} \in \text{“prefault” AND } k_{in} \in \text{“prefault” AND } k_{BB} \in \text{“prefault”} \\
\text{AND } k_{oil} \in \text{“prefault” AND } k_{LTC} \in \text{“prefault” AND } k_{cool} \in \text{“prefault”} \\
\text{AND } k_{tank} \in \text{“prefault” THEN} \\
k_{tot.resid.res} = a_{31} \cdot k_{wind.} + a_{32} \cdot k_{in.} + a_{33} \cdot k_{bush.} + a_{34} \cdot k_{oil} + a_{35} \cdot k_{LTC} + a_{36} \cdot k_{cool.} + a_{37} \cdot k_{tank} + c_3 \\
\text{IF } k_{wind} \in \text{“emergency” AND } k_{in} \in \text{“emergency” AND } k_{BB} \in \text{“emergency” AND } k_{oil} \in \text{“emergency”} \\
\text{AND } k_{LTC} \in \text{“emergency” AND } k_{cool} \in \text{“emergency” AND } k_{tank} \in \text{“emergency” THEN} \\
k_{tot.resid.res} = a_{41} \cdot k_{wind.} + a_{42} \cdot k_{in.} + a_{43} \cdot k_{bush.} + a_{44} \cdot k_{oil} + a_{45} \cdot k_{LTC} + a_{46} \cdot k_{cool.} + a_{47} \cdot k_{tank} + c_4 \\
\ldots
\end{cases} \tag{7.8}$$

For determination of the value of total residual resource coefficient of the transformer use of model inference Takagi-Sugeno. Model of logic equation is used. Mathematical model of total residual resource coefficient is the system of logic equations. Output of the model total $K_{tot.resid.res.}$ is found as weighted sum of conclusions (7.8) of rules basically written in the form the system of logic equations:

$$k_{tot.resid.res.} = \sum_{j2=1}^{m3} w_{j2} \begin{pmatrix} a_{j21} \cdot k_{wind.} + a_{j22} \cdot k_{in.} + a_{j23} \cdot k_{Bush.} + a_{j24} \cdot k_{oil} \\ + a_{j25} \cdot k_{LTC} + a_{j26} \cdot k_{cool.} + a_{j27} \cdot k_{tank} + c_{j2} \end{pmatrix}, \tag{7.9}$$

where: $0 \le w_{j_2} \le 1$ is the degree of execution (weight) of the j_2-th rule that is determined by the correspondence of real changes of diagnostic parameters of the transformer.

ANFIS is the simplest network of direct propagation that contains adaptive nodes, using the teaching rules the parameters of these nodes are arranged to minimize the error between the real output of the model $k_{tot.resid.mod.}$ and real total residual resource coefficient $k_{tot.resid.res}$ of the transformer:

$$\delta = \sqrt{\frac{1}{N_1}\sum_{k_3=0}^{N_1-1}(k_{tot.resid.res.modk3} - k_{tot.resid.res.k3})^2} \rightarrow \min \tag{7.10}$$

where: N – is number of rows in teaching sample; k_3 – is the number of the row in teaching sample, starting from the row with consecutive number "0".

Taking into account the iterative computation experiments carried out the vector of membership functions parameters is determined in Table 7.2.

Taking into account the data of the Table 7.1 and Table 7.2 and (7.9) we obtain mathematical model of the coefficient of total residual resource in the form:

Table 7.2. Parameters of membership function.

Parameters	Input parameters of the model	Name of the term	Number of the rule	Parameters of membership function	
				σ	C
Winding state	$K_{wind.}$	Normal	1	0.3825	0.7944
		Minor deviation	2	0.479	0.5197
		Prefault	3	0.4903	0.5668
		Emergency	4	0.4	0.1697
Insulation state	$k_{in.}$	Normal	1	0.3653	0.8698
		Minor deviation	2	0.4642	0.6104
		Prefault	3	0.5102	0.5267
		Emergency	4	0.3949	0.1742
State of BB	$k_{bush.}$	Normal	1	0.3202	0.9221
		Minor deviation	2	0.3419	0.7649
		Prefault	3	0.4914	0.5376
		Emergency	4	0.4032	0.1925
State of oil	K_{oil}	Normal	1	0.4369	0.9273
		Minor deviation	2	0.3404	0.9674
		Prefault	3	0.412	0.599
		Emergency	4	0.4031	0.2057
State LTC	K_{LTC}	Normal	1	0.3984	0.973
		Minor deviation	2	0.3316	0.963
		Prefault	3	0.4468	0.5881
		Emergency	4	0.4428	0.2349
State of cooling system	$k_{cool.}$	Normal	1	0.3439	1.153
		Minor deviation	2	0.3507	0.9706
		Prefault	3	0.437	0.597
		Emergency	4	0.4263	0.2397
State of tank	K_{tank}	Normal	1	0.3454	0.9506
		Minor deviation	2	0.3801	1.017
		Prefault	3	0.4582	0.6273
		Emergency	2	0.5451	0.564

$$
\begin{cases}
\text{IF } k_{wind.} \in \text{“normal” AND } k_{in} \in \text{“normal” AND } k_{BB} \in \text{“normal”} \\
\text{AND } k_{oil} \in \text{“normal” AND } k_{LTC} \in \text{“normal” AND } k_{cool} \in \text{“normal”} \\
\text{AND } k_{tank} \in \text{“normal” THEN} \\
k_{tot.resid.res} = 0{,}6166 \cdot k_{wind} + 0{,}4125 \cdot k_{in} + 0{,}4618 \cdot k_{BB} + 1{,}83 \cdot k_{oil} + 1{,}804 \cdot k_{LTC} \\
+ 0{,}0462 \cdot k_{cool.} + 1{,}96 \cdot k_{tank} - 5{,}377 \\
\text{IF } k_{tank} \in \text{“minor deviation” AND } k_{in.} \in \text{“minor deviation”} \\
\text{AND } k_{Bush} \in \text{“minor deviation” AND } k_{oil.} \in \text{“minor deviation”} \\
\text{AND } k_{LTC} \in \text{“minor deviation” AND } k_{cool.} \in \text{“minor deviation”} \\
\text{AND } k_{tank} \in \text{“minor deviation” THEN} \\
k_{tot.resid.res} = -0{,}0393 \cdot k_{wind.} + 0{,}2609 \cdot k_{in.} + 0{,}1086 \cdot k_{Bush.} - 0.37 \cdot k_{oil} - 0{,}1459 \cdot k_{LTC} \\
- 0{,}02387 \cdot k_{cool.} - 0{,}05863 \cdot k_{tank} + 0{,}1288 \\
\text{IF } k_{wind} \in \text{“prefault” AND } k_{in} \in \text{“prefault” AND } k_{Bush.} \in \text{“prefault”} \\
\text{AND } k_{oil} \in \text{“prefault” AND } k_{LTC} \in \text{“prefault” AND } k_{cool.} \in \text{“prefault”} \\
\text{AND } k_{tank} \in \text{“prefault” THEN} \\
k_{tot.resid.res} = -0{,}2165 \cdot k_{wind.} - 0{,}3174 \cdot k_{in.} - 0{,}4678 \cdot k_{Bush.} - 0{,}514 \cdot k_{oil} - 0{,}882 \cdot k_{LTC} \\
- 0{,}5302 \cdot k_{cool.} - 1{,}406 \cdot k_{tank} + 3{,}88 \\
\text{IF } k_{wind.} \in \text{“emergency” AND } k_{in} \in \text{“emergency” AND } k_{BB} \in \text{“emergency” AND } k_{oil} \in \text{“emergency”} \\
\text{AND } k_{LTC} \in \text{“emergency” AND } k_{cool.} \in \text{“emergency” AND } k_{tank} \in \text{“emergency” THEN} \\
k_{tot.resid.res} = 0{,}03166 \cdot k_{wind.} - 0{,}06144 \cdot k_{in.} - 0{,}387 \cdot k_{Bush.} + 0{,}06 \cdot k_{oil} + 0{,}3199 \cdot k_{LTC} \\
- 0{,}026 \cdot k_{cool.} - 0{,}006 \cdot k_{tank} + 0{,}003 \\
\ldots
\end{cases}
\quad (7.11)
$$

Fig. 7.2 shows that in the process of formation of the structure of neuro fuzzy model of the transformer seven inputs and one output of this model were set. Each of seven inputs has four terms. That is, each set of possible values of input parameters of the model is conventionally divided into four subset: “normal” values of input parameter, “minor deviations ” of the values of input parameter, and “prefault” values of input parameter, “emergency” values of input parameter. Membership degree of each value of input parameter to corresponding set of values is determined by Gaussian membership function.

The model is intended for determining the numerical value of total residual resource coefficient of the transformer, that is why it has one output. This numerical value is found by means of solution of linear equation, that describes the dependence of the coefficient of total residual resource of the transformer on input parameters.

The obtained neuro-fuzzy model allows us to determine the value of total residual resource coefficient of the transformer depending on the values of input parameters – residual resources coefficients by each of controlled diagnostic parameters. Thus, if each of residual resource coefficients of diagnostic parameters will equal 0 relative units (r.u.), then the coefficient of total residual resource (with the error 0.4%) equals 0.004 r.u.; if each of the coefficients of residual resource of diagnostic parameters equals 1. r.u., then the coefficient of total residual resource (with the error 3.2%) equals 0.968 r.u.

Despite the complexity of dependences, mathematical model of residual resource coefficient of the transformer (7.11) may be used for programming fuzzy controller in order to create the device for on-line determination of transformer state by means of analysis of residual resource coefficient of the transformer value.

7.3 ACCOUNT OF THE FORECAST CURRENT VALUE OF RESIDUAL RESOURCE OF THE TRANSFORMERS IN THE PROCESS OF OPTIMAL CONTROL OF MICRONETWORKS MODES

It is known that in the process of operation, energy enterprise plans to remove the equipment in the overhaul out of service, cost of is forecast. Removal of the transformer into overhaul in a planned number of years (T_{WF}) of trouble-free operation (12 years) provides certain list of works and their expected cost $B_{oh\ pl}$. For instance, for 330/110 kV transformers of 125–250 MVA power the cost (B) of such repair amounts to 200–300 thousand UAH. We propose to assume that removal the transformers out of service into current repair requires unscheduled cost.

The cost of repair may increase by the value ΔB_1, replacement of damaged blocks of the transformer and additional work, connected with the replacement. These costs are not provided in case of "typical" planned overhaul:

$$\Delta B_1 = \sum_{i=1}^{n} \left(B_i \cdot e^{\gamma_i \cdot k_{res.i}^{\beta_i}} \right), \tag{7.12}$$

where: B_i – is the cost of replacement of ith damaged block of the transformer and additional work, connected with this replacement, n – is a number of damaged blocks that require unscheduled replacement; $k_{res.i}^{\beta_i}$ – is residual resource coefficient of i-th block that requires unscheduled replacement; γ and β – are coefficients, that characterize the impact of residual resource coefficient on the expected cost of unscheduled repair or replacement of ith block of the transformer (is determined by means of processing of statistic data).

Repair cost may increase by the cost of ΔB_2 (as compared with expected) in case of enlarged current (instead of planned overhaul) repair of the transformer, that did not operate for planned number of years:

$$\Delta B_{2_j} = \left(1 - e^{\alpha_j (T_j - 1)}\right) \cdot B_{OH}, \tag{7.13}$$

where j – is a number of the transformer, T_j – is time, the ith transformer functioned after putting into operation or after the last overhaul (enlarged current) repaint to the moment of mode control, λ – is the coefficient, that characterizes the intensity of cost growth ΔB_2 that depends on the construction of the transformer, conductions and operation mode (is determined experimentally), B_{OH} – is the cost of transformer overhaul.

It should be noted that removal the transformer out of service takes place not only as a result of relay protection, emergencies control automation operation, but also by a person responsible for safety operation by the results of control of diagnostic parameter, values of which sometimes only approaches to limiting values.

Within the context of creation of modern Smart Grids and to provide safe, reliable, quality and economic efficient operation of EES it is necessary to perform the control over active power overflow to realize by means of the transformer, performing reliable and information archons on the mode. That is why we suggest to take into account the coefficient of regulating transformer limitation:

$$k_{wind,j} = (1 - k_{res,j}) \cdot B_{cq,j} \tag{7.14}$$

where: B_{cq} – is the coefficient of repair cost value growth of the jth transformer.

$$B_{cq,j} = \frac{\Delta B_{1,j} + \Delta B_{2,j}}{\Delta B_{1,j} + B_{pl.,j}}. \tag{7.15}$$

As the example, we will consider 24 nodes test circuit, recommended by IEEE (Fig. 7.3). In branches 11–9, 12–14, 12–9, 11–4, and 3–7 transformers АТДЦТН-63000/230/138, АТДЦТН-100000/230/138, АТДЦТН-125000/230/138 are installed. Initial node loads, optimal complex transformation ratios and corresponding transformers LTC positions (number of taps) are given in Tables 7.3 and 7.4.

Table 7.3. Information of circuit branches.

Branches					
Number of beginning	Number of the end	R, Ohm	X, Ohm	k_{active}	$k_{reactive}$
11	10	0.6	27	0.6487	0
12	9	0.37	9.28	0.6498	0
11	9	0.3	13	0.6479	0
7	3	0.21	11.53	0.65	0
1	2	0.4951	2.6471	1	0
1	3	10.398	40.221	1	0
1	5	4.1516	16.092	1	0
2	4	6.2464	24.129	1	0
2	6	9.4649	36.565	1	0
3	9	9.882	20.962	1	0
4	9	5.1038	19.749	1	0
5	10	4.342	16.816	1	0
8	9	8.82	15.124	1	0
8	10	2.067	2.145	1	0
11	23	5.207	22.793	1	0
11	14	2.8566	22.112	1	0
12	23	5.207	25.18	1	0
12	13	6.5596	51.101	1	0
23	13	5.8719	45.759	1	0
14	16	2.645	20.578	1	0
15	16	2.338	8.404	1	0
16	17	1.7457	13.701	1	0
16	19	3.117	11.206	1	0
17	18	0.9522	7.6176	1	0
17	22	7.1415	55.704	1	0
21	22	4.6023	35.866	1	0
7	15	5.68	24.865	1	0
21	18	0.873	6.851	1	0
21	15	1.666	12.96	1	0
19	20	1.349	10.474	1	0
20	13	0.741	5.713	1	0
12	10	0.31	14	0.6524	0
10	6	2.6471	11.522	1	0

In order to simplify the understanding of the results obtained (by means of equivalenting) the loading parameters of 7th node to take into account in 8th node and change number of 24 node for number 7.

Having known the circuit and normal node parameters, we define optimal transformation ratios:

$$\mathbf{k}_{\mathbf{a.opt}} = 1 - \mathrm{diag}\left(\mathrm{Re}\left(-\mathbf{N}_{\mathbf{k.bal.}_b} \cdot \mathbf{Z} \cdot \mathbf{C}_e \cdot \mathbf{J}\right)\right) \cdot \mathbf{U}_b^{-1} \cdot \mathbf{E}^*_{\mathbf{bal.}_a} \tag{7.16}$$

$$\mathbf{k}_{\mathbf{r.opt}} = -\mathrm{diag}\left(\mathrm{Im}\left(-\mathbf{N}_{\mathbf{k.bal.}_b} \cdot \mathbf{Z} \cdot \mathbf{C}_e \cdot \mathbf{J}\right)\right) \cdot \mathbf{U}_b^{-1} \cdot \mathbf{E}^*_{\mathbf{bal.}_r} \tag{7.17}$$

Table 7.4. Information of the circuit nodes.

The node №	U, kV	Phase, grad	Pload, MW	Qload, MVAr	Pgen, MW	Qgen, MW
1	136.34	–5.73	108	22	182	30
2	135.54	–6.27	187.15	76	172.9	30
3	141.22	–0.94	176.4	36.26	0	0
4	136	–8.2	74	15	0	0
5	137	–8.11	68.16	13.44	0	0
6	136.36	–10.3	129.2	25.27	0	0
7	218.84	2.99	19.4	1.94	0	0
8	142.57	–7.72	169.29	34.65	131.55	131.75
9	141.55	–5.6	275	66	0	0
10	141.64	–7.29	191.1	49	0	0
11	222.36	–3.12	40	10	0	0
12	219.92	–4.12	54.88	17.64	0	0
13	236.15	7.89	0	0	495	150
14	228.34	1.56	184.3	37.05	0	101.39
15	233.22	11.21	304.32	61.44	235.2	51.32
16	233.45	10	100	20	185	80
17	237.79	15.28	35.64	13.86	0	0
18	241.5	16.92	323.01	65.96	417.1	176.19
19	231.47	7.6	177.38	36.26	0	0
20	233.81	7.31	128	26	0	0
21	241.5	17.83	0	0	425.7	146
22	241.5	25.95	0	0	420	–3.69
23	234.6	0	265	54	417.85	281.37

where $\mathbf{N}_{k.bal.b}$ – is the second matrix of branches connection in contour balanced transformation ratios; $\mathbf{Z}$ – diagonal matrix of complex branches resistances; $\mathbf{C}_e$ – is the matrix of currents distribution coefficients for economic mode of electric network (corresponds to minimal losses of electric energy); $\mathbf{J}$ – vector-column of currents in nodes; $\mathbf{U}_b$ – is the voltage of basic node; $E^*_{bal.a}$, $E^*_{bal.r}$ – are optimal balancing electric moving in relative units (EMF) (active and reactive components).

Taking in the consideration the discrete character of LTC switching's, errors of instrument transforms, errors of data transmission channels and recommendations (Tirupathi et al. 2012), we assume that non-sensitivity zone of active power losses may be considered as regulating actions on LTC of the transformer – to 3% (Constantin 2013).

As initial conditions, we assume that, in accordance with load graph, LTC of transformers 9–11, 9–12, 11–10, 12–10 have transformation ratios 0.6413 (14 tap), 0.6347 (14 tap), 0.6397 (14 tap), 06446 (14 tap). It should be noted that further changes of operation mode were realized at admissible voltage deviations ±5%, from nominal voltage U_{nom}. Regulation of the transformer 7–3 is inexpedient on conditions of the usage of the given technique of determination of optimal transformation ratios.

We define the losses of active and reactive power in the branches of the circuit at current transformation ratios (Table 7.5).

$$\Delta S_{\Sigma br} = \Delta P_{\Sigma br} + j\Delta Q_{\Sigma br} = 3 \cdot \sum_{j=1}^{m} \Delta S_{br,j}, \qquad (7.18)$$

where $\Delta \mathbf{S}_{br} = \mathrm{diag}\,(\Delta \mathbf{U}_{br}) \cdot \hat{\mathbf{I}}_{br}$ – is vector-column of complete power losses in the branches of the circuit. $\mathbf{E}_{br}$ – is vector column of the current in branches, m – is a number of the branch

Table 7.5. Parameters of the current mode.

	The node number																						
P-ters*	1	2	3	4	5	6	7	8	9	10	11	12	13	14	15	16	17	18	19	20	21	22	23
Pload, MW	19.44	37.43	38.8	17.02	13.63	24.55	3.49	27.08	60.5	38.22	8.4	10.43	0	29.49	39.6	20	8.55	80.75	35.47	26,88	0	0	53
Qload, MVAr	3.96	15.2	7.97	3.45	2.69	4.8	0.35	5.54	14.52	9.8	2.1	3.35	0	5.93	8	4	3.33	16.49	7.25	5,46	0	0	10,8
ΔPΣ, MW	4.49																						
ΔQΣ, MVAr	29.05																						
11–10										0.6437													
Tap										**14**													
12–10										0.6446													
Tap										**14**													
9–11									0.6413														
Tap									**14**														
9–12									0.6397														
Tap									**13**														
7–3			0.65																				
Tap			**14**																				

* – parameters.

in the circuit, $\Delta \mathbf{U}_{br} = \mathbf{M}_{\Sigma} \cdot \mathbf{U}_{node}$ – is vector column of phase voltages in the nodes, $\mathbf{\Delta P}_{br}$, $\mathbf{\Delta Q}_{br}$ – is vector-column of active and reactive power losses in the branches of the circuit (correspondingly).

We define optimal transformation ratios (7.16–7.17) and position of LTC on condition of minimal amount of switchings (in order to maintain switching resources of LTC) to provide minimal losses of active power in branches of the circuit in Table 7.6.

As a result of performance of control actions, mode optimization power losses were reduced from $\Delta S_1 = 4.49 + j29.05$ (MVA) for mode (Table 5) to $\Delta S_2 = 4.42 + j28.69$ (MVA). Thus, the effect of transformer LTC switchings is ΔS_1-$\Delta S_2 = 0.07 + j0.36$ (MVA). The transition from the current to the optimal mode may be performed by switching the LTC of the transformer, installed in the branch 9–12 from 13 tap to 14 and carry out transformer regulation changing position of LTC from 14 tap to 15. We consider the transition to other stage of the daily load graph, which parameters are given in Table 7.7.

As a result of performing of the control actions, we achieved the mode optimization by decreasing power losses from $\Delta S_1 = 61.85 + j412.73$ (MVA) for the mode (Table 7.7) to $\Delta S_2 = 61.80 + j412.68$ (MVA). Thus, the effect of LTC transformer switchings is ΔS_1-$\Delta S_2 = 0.05 + j0.5$ (MVA).

If, as a result of determining of the coefficient of regulating the effect limitation for circuit transformers (Fig. 7.3) the following values are obtained: $k_{wind,9-11} = 0.85$, $k_{wind,12-9} = 0.4$, $k_{wind,11-10} = 0.3$ a $k_{wind,12-10} = 0,2$, then expected quasi-decrease of losses, taking into account these coefficient, will be defined.

Control actions are performed by the transformer, installed in the branch 9–11, namely we change position of LTC with 14 tap on 15. In this case, the expected losses of active power are $\Delta P_{9-11} = 61.82$ (MW). We find the decrease of active power losses $\Delta P_{\Sigma} - \Delta P_{9-11} = 61.85 - 61.82 = 0.03$ (MW), however, having taken into account the coefficient of regulating effect limitation, losses decrease change $\delta P_{quasi.9-11} = (\Delta P_{\Sigma} - \Delta P_{9-11}) \cdot k_{wind,9-11} = 0.0255$ (MW). New quasi-losses $\Delta P_{quasi.9-11} = \Delta P_{9-11} + \delta P_{quasi.9-11} = 61.82 + 0.0255 = 61.8455$ (MW). Results of the calculation of other transformers are given in Table 7.9.

Table 7.6. Optimal transformation ratios for the current mode.

	The node number																						
P-ters	1	2	3	4	5	6	7	8	9	10	11	12	13	14	15	16	17	18	19	20	21	22	23
Pload, MW	19.44	37.43	38.8	17.02	13.63	24.55	3.49	27.08	60.5	38.22	8.4	10.43	0	29.49	39.6	20	8.55	80.75	35.47	26.88	0	0	53
Qload, MVAr	3.96	15.2	7.97	3.45	2.69	4.8	0.35	5.54	14.52	9.8	2.1	3.35	0	5.93	8	4	3.33	16.49	7.25	5.46	0	0	10,8
ΔPΣ, MW	4.42																						
ΔQΣ, MVAr	28.69																						
11–10										0.6513													
Tap										14													
12–10										0.6542													
Tap										15													
9–11									0.6507														
Tap									14														
9–12									0.6521														
Tap									14														
7–3			0.65																				
Tap			14																				

Table 7.7. Parameters of the mode after load change.

	The node number																						
P-ters	1	2	3	4	5	6	7	8	9	10	11	12	13	14	15	16	17	18	19	20	21	22	23
Pload, MW	108	187.1	176.4	74	68.16	129.2	19.4	169.2	275	39.1	40	54.88	0	184.3	304.3	100	35.64	323.0	177.3	128	0	0	265
Qload, MVAr	22	76	36.26	15	13.44	25.27	1.94	34.65	66	10	10	17.64	0	37.05	61.44	20	13.86	65.96	36.26	26	0	0	54
ΔPΣ, MW	61.85																						
ΔQΣ, MVAr	412.73																						
11–10										0.6513													
Tap										14													
12–10										0.6542													
Tap										15													
9–11									0.6507														
Tap									14														
9–12									0.6521														
Tap									14														
7–3			0.65																				
Tap			14																				

We find losses of active power in the αth branch, that contains the transformer, as function the αth element of vector-column of complete power losses in the branches of the circuit by the expression:

$$\Delta P_{\alpha} = \mathrm{Re}(\Delta S_{\alpha}), \tag{7.19}$$

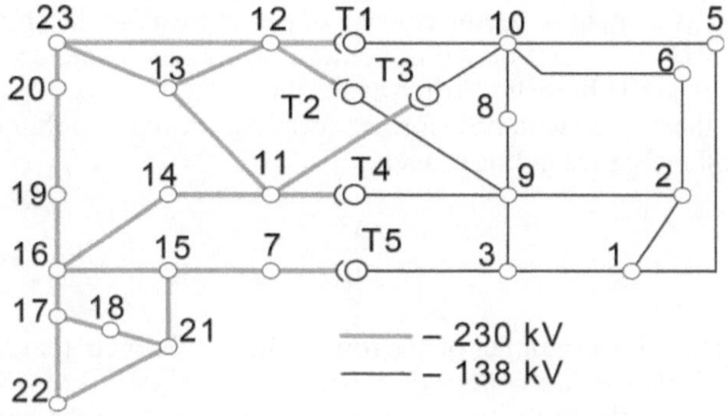

Figure 7.3. Test scheme IEEE 24 nodes.

Table 7.8. Parameters of the optimal mode without taking into account the technical state of the transformers.

	The node number																						
P-ters	1	2	3	4	5	6	7	8	9	10	11	12	13	14	15	16	17	18	19	20	21	22	23
Pload, MW	108	187.1	176.4	74	68.16	129.2	19.4	169.2	275	39.1	40	54.88	0	184.3	304.3	100	35.64	323.0	177.3	128	0	0	265
Qload, MVAr	22	76	36.26	15	13.44	25.27	1.94	34.65	66	10	10	17.64	0	37.05	61.44	20	13.86	65.96	36.26	26	0	0	54
ΔPΣ, MW	61.80																						
ΔQΣ, MVAr	412.68																						
11–10										0.665													
Tap										16													
12–10										0.651													
Tap										14													
9–11									0.659														
Tap									15														
9–12									0.6419														
Tap									13														
7–3			0.65																				
Tap			14																				

Table 7.9. Results of limiting effect coefficients calculation for circuit transformers.

Transformer	$K_{tr.cur.}$ $N_{cur.}$	$K_{tr.opt.}$ $N_{opt.}$	$K_{wind.j}$	$\Delta P_{tr.j}$ MW	$\Delta P_{\Sigma} - \Delta P_{tr.j}$ MW	$\delta P_{quasi.j}$ MW	$\Delta P_{quasi.j}$ MW
9–11	0.6507 14	0.659 15	0.85	61.82	0.03	0.0255	61.8455
12–9	0.6521 14	0.6419 13	0.46	61.83	0.02	0.0092	61.8392
11–10	0.6513 14	0.665 16	0.34	61.835	0.015	0.0051	61.8401
12–10	0.6542 15	0.651 14	0.25	61.84	0.01	0.0025	61.8425

where: $\Delta S_{\alpha} = \Delta U_{\alpha} I_{\alpha}$ – is element of vector-column of power losses in the branches that contain transformers, ΔU_{α} – is kth element of vector-column of phase voltages drop in the branches, and I_{α} – is the current of the branches with transformers couplings, α – is the number of row that corresponds to the branch with transformer couplings in vector-column ΔS_{br}.

The value of quasi resistance in kth-branch:

$$Z_{\alpha} = \frac{\Delta S_{\alpha}}{\hat{I}_{\alpha}^{2}}, \tag{7.20}$$

where $\alpha = k+\beta$, where k – is the number of the row of the first branch, that contains the transformer, β – is the coefficient of the change of consecutive number of branch that contains the transformer; it changes in the range from 0 to $(\psi - 1)$, where ψ – is the number of branches containing transformers.

Applying this algorithm, according to (7.20), quasi-resistances of the branches containing transformers are found. The results of the calculations are given in Table 7.10.

We define mode parameters for the circuit with quasi-resistances from the Table 7.11 and optimal corrected transformation ratios from Table 7.12.

The aim of the optimal control is to provide the minimum of all system active power losses that are determined by the expression:

Table 7.10. Quasi-resistances of transformers branches of the circuit.

Parameters	Transformer 9–11	Transformer 12–9	Transformer 11–10	Transformer 12–10
Branch resistance, Ohm	0.3 + j13	0.37 + j9.28	0.3 + j27	0.3 + j14
Quasi resistance of the branch, Ohm	0.32 + j25.2	0.4 + j13.2	0.36 + j28.4	0.35 + j19.2

Table 7.11. Parameters of normal mode after loads change, taking into account the technical state of the transformers.

	The node №																						
P-ters*	1	2	3	4	5	6	7	8	9	10	11	12	13	14	15	16	17	18	19	20	21	22	23
Pload, MW	108	187.1	176.4	74	68.16	129.2	19.4	169.2	275	39.1	40	54.88	0	184.3	304.3	100	35.64	323.0	177.3	128	0	0	265
Qload, MVAr	22	76	36.26	15	13.44	25.27	1.94	34.65	66	10	10	17.64	0	37.05	61.44	20	13.86	65.96	36.26	26	0	0	54
ΔPΣ, MW	62.47																						
ΔQΣ, MVAr	422.16																						
11–10										0.665													
Tap №										16													
12–10										0.651													
Tap №										14													
9–11									0.659														
Tap №									15														
9–12									0.6419														
Tap №									13														
7–3			0.65																				
Tap №			14																				

* – parameters.

Table 7.12. Parameters of the optimal mode after the load change, taking into account the technical state of the transformers and corrected transformation ratios.

	The node №																						
P-ters	1	2	3	4	5	6	7	8	9	10	11	12	13	14	15	16	17	18	19	20	21	22	23
Pload, MW	108	187.1	176.4	74	68.16	129.2	19.4	169.2	275	39.1	40	54.88	0	184.3	304.3	100	35.64	323.0	177.3	128	0	0	265
Qload, MVAr	22	76	36.26	15	13.44	25.27	1.94	34.65	66	10	10	17.64	0	37.05	61.44	20	13.86	65.96	36.26	26	0	0	54
ΔPΣ, MW	62.46																						
ΔQΣ, MVAr	422.28																						
ΔPΣ, MW										0.664													
ΔQΣ, MVAr										15													
11–10										0.641													
Tap №										13													
12–10									0.6753														
Tap №									17														
9–11									0.6492														
Tap №									14														
9–12			0.65																				
Tap №			14																				

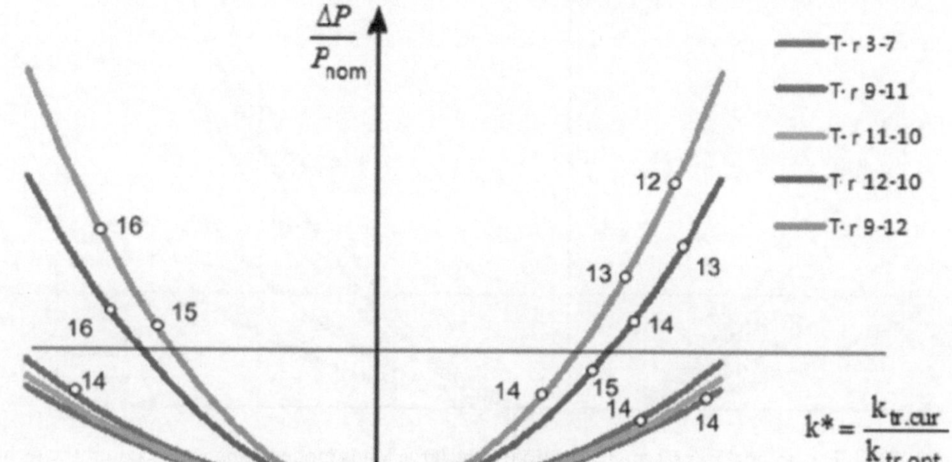

Figure 7.4. The optimal transformation ratios for the small loads mode.

$$\Delta F = \sum_{i=1}^{n} \Delta P_i \rightarrow \min. \qquad (7.21)$$

If ΔF_{min} – is the minimum value of the efficiency function, n – is the total number of branches in the circuits, and $k_{tr.opt}$ – is the optimal value of the transformation ratio, then we build the dependence of active power losses change $\Delta F^* = \frac{\Delta F}{\Delta F_{min}}$ on the transformation ratios $k^* = \frac{k_{tr.cur}}{k_{tr.opt}}$ (Fig. 7.4).

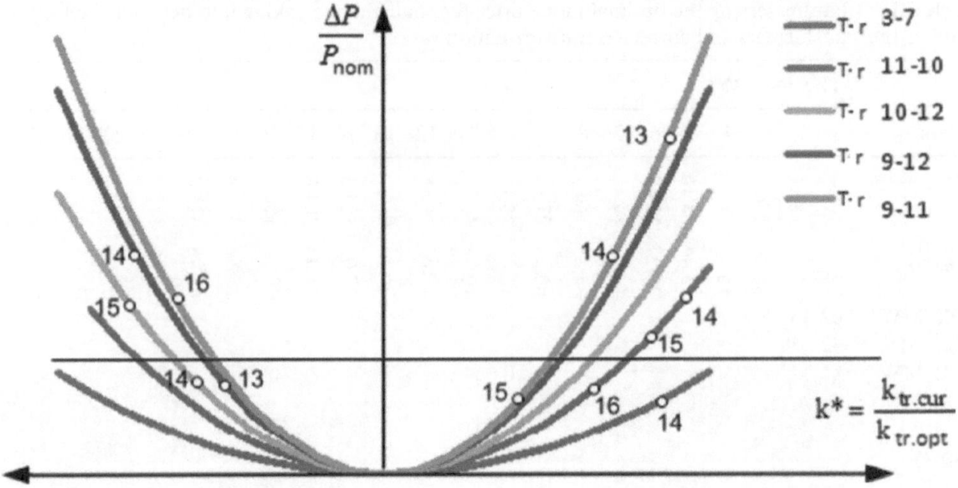

Figure 7.5. The optimal transformation ratios for the large loads mode without taking into account the technical state of the transformers.

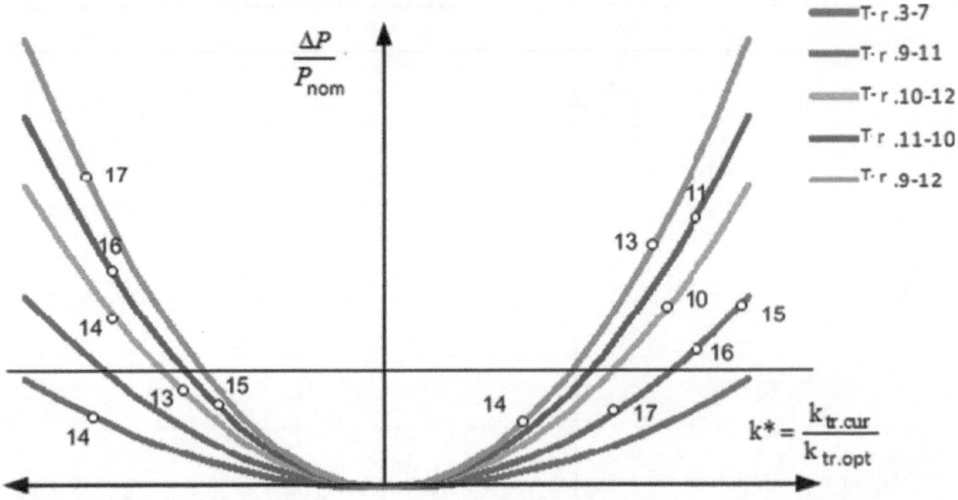

Figure 7.6. The optimal transformation ratio for the large loads mode taking into account the technical state of the transformers.

Thus, the transition from the current to the optimal mode can be performed by switching the LTC of the transformer installed in the branch 9–12 from 13 tap to 14 tap and regulation of the transformer in branch 12–10, changing the LTC position from tap 14 to 15 tap.

As a result of performance of the control actions, the optimal mode will be reached by switching the transformer in the branch 11–10 from 14 to 16 tap of the LTC, the transformer in the branch 12–10 from 15 tap to 14 tap, the transformer in the branch 9–11 from 14 to 15 tap and the transformer in the branch 9–12 from 14 to 13 tap of the LTC, respectively.

We see that due to taking into account the technical state of the transformers, their ranking occurred by the measure of impact on the reduction of the active power losses. To reach the optimal mode, it is more expedient now to use a transformer of 9–12 branch, as it reduces most active power losses during one switching of the LTC from 13 tap to 14 tap.

7.4 CONCLUSIONS

1. Application of the methods of the neuro-fuzzy modeling allows to calculate the value of the residual resource coefficient of the power transformer, taking into consideration both current values of certain diagnostic parameters as retrospective values.
2. Analysis of calculation values of the residual resource coefficient of the power transformer allows to evaluate its current technical state.
3. Method of determination of optimal control actions by the transformers with the LTC is improved, taking into consideration their technical state. It is expected that application of this method enables to evaluate the expediency of performance of the control actions.
4. It is determined that performance of such optimal transformation ratios will lead to decrease in active power losses and excessive expenses for transformers repair in case of their failure.

REFERENCES

Alekseev B.A., 2010: Large power transformers: state control in the process of operation and revision [Text] / Moscow NTF «Energoprogress», pp. 88.

Bocovich M., Iyer K., Terhaar R.M., Mohan N., 2013: Overview of series connected flexible AC transmission systems (FACTS), North American Power Symposium (NAPS), pp. 246–249.

Buslavets O.V., Lezhniuk P.D., Rubanenko O.E., 2015: Evaluation and increase of load capacity of on-load tap changing transformers for improvement of their regulating possibilities, Eastern-European Journal of Enterprise Technologies. Vol. 2/8 (74), pp. 35–41.

Constantin C., 2013: Power flow control solutions in the Romanian power system under high wind generation conditions/C. Constantin, M. Eremia, L. Toma // PowerTech (POWERTECH), IEEE Grenoble, pp. 1–6.

Jakushokas, R., Friedman, E.G., 2013: Power Network Optimization Based on Link Breaking Methodology, IEEE Transactions on Very Large Scale Integration (VLSI) Systems. Vol. 21, Issue 5, pp. 983–987.

Kolcun M., Hlubeň D., Beňa L., Djagarov N., Grozdev Z., 2010: Transformer use for active power flow control in the electric power system, Environment and Electrical Engineering (EEEIC), 9th International Conference, pp. 246–249.

Kylymchuk A.B., Lezhnyuk P.B., Rubanenko O.E., 2015: Reduction of Additional Losses of Electric Energy in Parallel Operating Non-Uniform Electrical Grids Taking into Account Non-Uniformity and Sensitivity, International Journal of Energy Policy and Management. Vol. 1, pp. 1–5. – Mode of access: World Wide Web: http://article.aascit.org/file/html/8950726.html.

Lezhnyuk P.D., Rubanenko O.E., Nikitorovych O.V., 2012: On-line diagnostics of high voltage equipment in problems of the optimal control of electric power systems modes. Technical electrodynamics. Vol. 3, pp. 35–36.

Rassalsky A.M., 2010: Integrated approach to the diagnostics of high-voltage substation equipment 220–1150 kW under operating voltage in operation conditions [Text] / A.M. Rassalsky, A.A. Sahno, S.P. Konogray, A.A. Hooke // Electric engineering and electro mechanics. Vol. 4, pp. 23–25.

Stogniy B.S., Kirilenko O.V., Butkevich O.F., Sopel M.F., 2009: Application of transient modes monitoring facilities in UES of Ukraine for the solution of the problems of a supervisory control, Procedures of the Institute of electrodynamics of national academy of sciences of Ukraine/Collection of scientific works. – K.: IED NASU, Issue 23, pp. 147–155.

Stogniy B.S., Sopel M.F., 2013: Fundamental principles of monitoring in power industry. About the action of monitoring // Technical electrodynamics. No.1., pp. 62–69, Refs.: 8.

Tirupathi R., Aruna G., Khan M.I., Ramesh K., 2012: Application of Phase Shifting Transformer in Indian Power System, International Journal of Computer and Electrical Engineering. Vol. 4, Issue 2, pp. 242–245.

Author index